HAMMERING INTO THE FUTURE

One Blacksmith Legacy

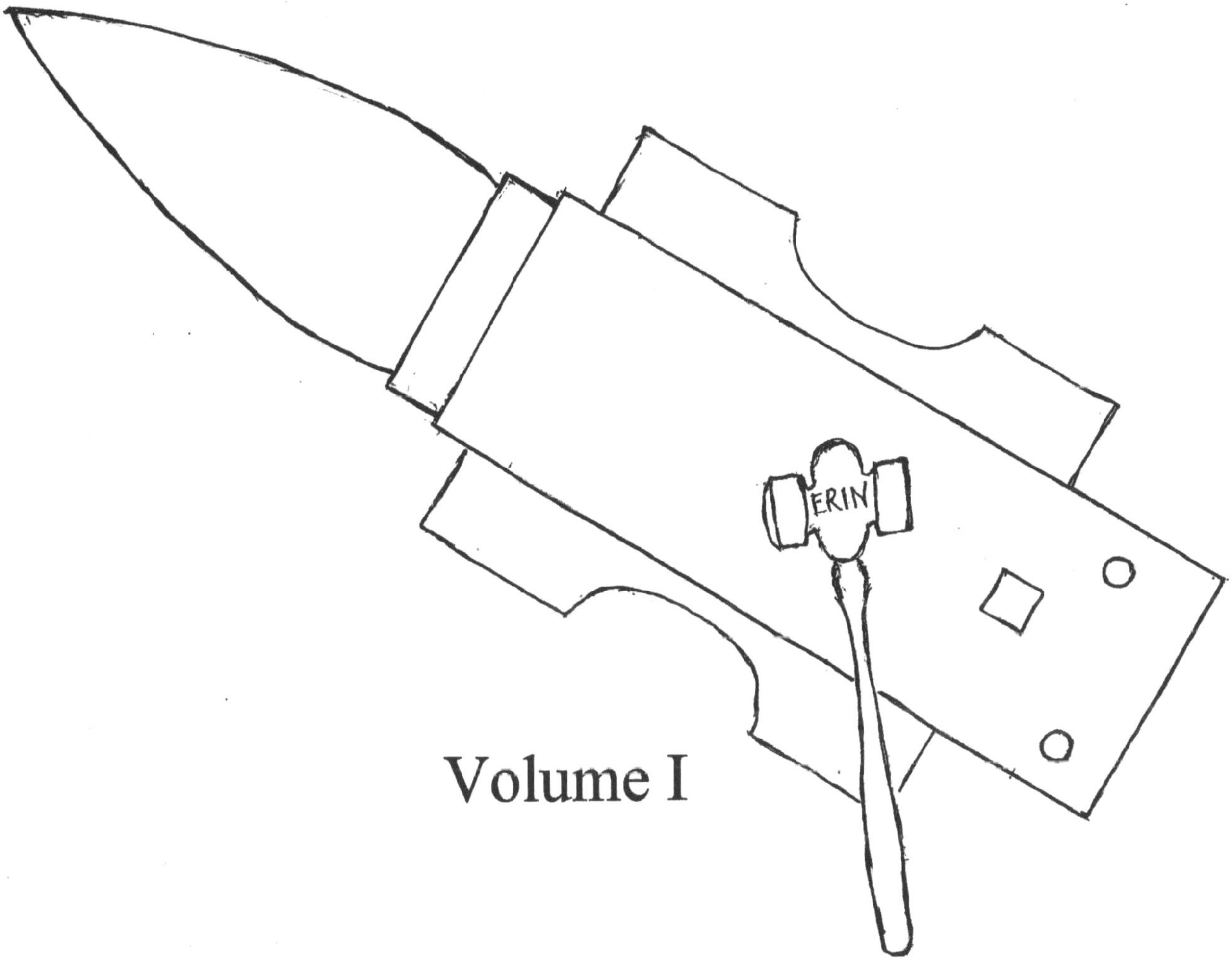

Volume I

Written

by

Erin A. Simmons

First Printing 2005
P.O. Box 1655
Shingle Springs California 95682
Photographs and drawing by: Erin A. Simmons

Library of Congress
ISBN: 9780976273912

FOREWORD

After many years of practice, workshop clinics, and travel to innumerable blacksmith operations we felt that the private notes should be published. *Hammering Into the Future* encompasses information and projects which are suitable for blacksmiths who are beginning their careers and seek solid guidelines, as well as a fund of experience to draw upon. The more complex tooling used regularly for our shop will challenge the experienced minds and hands which desire and thirst for projects to help streamline their own shop desires.

DEDICATION

I would like to dedicate this publication to those who have given unselfishly of their patience during the compilation of my shop noted to book form.

First for my wife Dusty, who helped open the road and smooth out the bumps. Next to Brian Bell of Wellington, Nevada for his friendship, good student aptitude and editorial comments. To Bill Curry of Placerville, California for his technical prowess and steadfast ideals. To Shayne Carter of Draper, Utah who came into our lives with knowledge and planted the seed we enriched with skill to make the "Tool Daddy" status exist. To Dennis Manning of Utah, the first blacksmith I could recognize as a true handmade tool artisan, where form, utility, and function met beauty. To Gene Armstrong and Gail without whom life may have taken a much different career course. Honesty and integrity always paramount for us and the animal. Fills my soul. To Michael and Elizabeth Easterling for their technical and editorial prowess, without which this publication would be less colorful. To family, friends, secretaries and countless students who in any and every way provided encouragement and asked the right questions. For those who demanded some ideas and experiences be put on paper before they were forgotten, this is for you.

DISCLAIMER

The information listed herein is an attempt to provide solid guidelines for forging skills in iron and steel.

The techniques have all been tested and tried with a great degree of success. Sound results have been possible and repeatable in succession through careful practice and procedure.

The ideas conveyed are to be considered as sound suggestions to help the student further their background and understanding of the craft without injury. Caution and care should always be foremost in the smith's mind. Whether working with hot or cold steel, sharp or blunt tools can cause serious injury.

No responsibility will be accepted for personal or patron injury as a result of proper or improper execution of the lessons provided herein. Proper etiquette is a relative term for participant safety, yet there are standards to which we must all adhere for no bodily harm to occur. Material destruction can occur even with due diligent attention and no fault of our own.

SAFETY

Protective gear and clothing are strongly suggested. Good fitting safety glasses for eyes are worn continuously. Ear plugs and protective gloves for our hands. Gauntlets on the gloves and boots are common place to prevent slag/spark burns. Hair and ears can be covered by a ball cap or bandana and/or hard hat. Loose fitting clothes should be comfortable but not shabby; no loose fabric is acceptable. Leather or Kevlar bib aprons are helpful and less flammable. Non restrictive attire is a must, but never sloppy. Tattered shirt tails can easily be ignited from a flame - or worse. Open shirt cuffs can easily catch in a bench grinder causing mortal destruction.

Use the utmost in common sense!

Fire extinguishers placed strategically within reach are a necessity. Keeping a slack/quench water tub full in the shop work area is of paramount importance. This may save the shop or its operator if/when an emergency arises.

Caution should continuously be exercised whether the process is a new or first time procedure, or has been executed many times by a seasoned blacksmith. We are all human and will only get so many close calls. Be alert, and take baby steps until all becomes familiar. Only then, can projects become a routine activity. Both eyes and ears are wide open and focused. Our mind is thoroughly engaged without distractions.

The risk that you accept can be minimal or maximal. The ground rules set for each of us will dictate who leaves unscathed. The dangers exist in each of us as an operator. Proceed at your own risk. Be careful and be safe.

MISSION STATEMENT

(Food for Thought)

The purpose of this publication is to stimulate the modern day blacksmith's imagination to provide a wider array of skills and venue to hone them.

To help the blacksmith develop a clearer mental picture of their project and pick the forging techniques which will bring that current idea to life.

To help develop for the smith a healthy repertoire and potential arsenal of approaches for blacksmithing.

To aid in presenting an understanding of possible and reasonable approaches while helping the smith understand other methods which are less probable.

We will find ways to make our individual forging techniques reliable and dependable.

We will shed light on methods which enable reasonable achievement, where to start and when to call it finished.

Discovering many of the benefits and few short comings of forging to utilize the elastic properties of hot steel.

We will endeavor to disclose what can and cannot be done with the blacksmith, a hand hammer and anvil and when its time to go to the power hammer.

How to select a reasonable volume of steel appropriate for a project.

We will discover ways in which hand hammer force verses depth of impact can be rated. Is the hammer too big or too small? Must the point of impact be smaller or larger? Is all else well, but the angle of billet approach skewed?

To revisit some of the old tried and tested methods and help the smith in understanding why they work consistently.

Which items should be single run forgings and which can be set for mass production?

To help visualize which tools will perform what task and building them proportionately.

To provide a reference for the smith, whether beginning or master, to disclose but some of our tooling and its uses.

We will uncover methods and trends of thought in which to feel our way through many projects whether a first attempt or old hat.

About the Author

Erin Adano Simmons

- A life long advocate of finding uses for steel which has lost its way.
- Professional horseman of 23 years.
- Farrier and blacksmith of 21 years
- Classical violin player
- Welder: Ford Motor Company, Mah Wah Plant, NJ
- Wilderness Ranger and Farrier of U.S. Forest Service, Willowa Whitman, Oregon.
- Seven years of university study:
  - Rutgers University, New Brunswick, NJ
  - Cook College - Plant Science
- California Polytechnic State University, San Luis Obispo, CA
- Graduate, School of Farrier Science, 1984

  Courses of concentration:
  - Crop Science
  - Preliminary Veterinary Medicine
  - Equine Science
  - Welding Technology

Table of Contents

Table of Contents (Continued)

Table of Contents (Continued)

TYPES OF BLOWS – HAND HAMMER SWINGS

Flat	
Angled Edge	Finger Knuckle: Excellent Hammer Heel (near): Efficient Hammer Toe (far): Less efficient
Brush Stroke	At impact: Away from smith
Short/Long: Stroke	Towards the smith – the smith's arm gauging power requirements
Push	As a drop hammer falls (rivet setting stroke)
Drawing	Vertical swing yet pushing material way from smith at impact. Full arm stroke, gracefully loose wrist for whip.
Hardy Hole	As a bridge collapses, so does the steel spanning the hardy hole.
Pinching	1/4-1/3 of hammer face driven into work on the anvil.
Lean	Striker resting sledge on work, the smith hammering just past an anvil edge.

TYPES OF HAND HAMMER BLOWS

Types of hand hammer blows

Full Face	Flat, rounding, pein (ball, cross or straight) Brush stroke (often on and past far edge of anvil, heel of anvil) Upset (into center face of anvil, hammer face larger than work piece beneath) Above anvil waist Above anvil tail
Half Face (to anvil)	Near edge of anvil Far edge of anvil Knuckle side edge (of hammer) Thumb side edge (of hammer) Heel blow (of hammer) Far edge (toe of hammer) Into step Across step a long step
Impact	Straight pein (full face, near and far edge, half face, heel (off side edge), into step of anvil, across step of anvil)
Horn	Small radius all the way up to larger radius Offsets and cross peins (heel of anvil, full face, into near and far edges, off near and far edges, across step of anvil, into step of anvil)
Over holes	Hardy Pritchel
Swing	Hammer handle length Our body configuration Arm length Torso considerations The heavier the hammer the shorter the handle

Most all hammers can be used for drawing and straightening to some extent. The best idea though is to use the right hammer for the job…style and heft is crucial. Each smith will have an appropriate weight and style which works best for their body size and arm strength. Hammer handle length and diameter are also paramount considerations. These hammer blows in singularity and in combination will create untold shapes for the smith. They are repeatable and dependable when practiced to perfection. With attention to detail and common sense, other useful types of blows may be discovered. Remember that between the swinging hand hammer and anvil or die beneath, the steel should be hot and resting!

Picture 1

Shown is a 400 pound anvil. We are looking at a tail to horn end view. Successive pictures of this chapter's views are of the same anvil for clarity. The following views are to depict the types of hammer blows possible by hand hammer and anvil. Hot steel is placed and held between the hammer and anvil before any attempt to forge is initiated. The hot steel is but resting on the anvil and the hand hammer is then set in motion to achieve the desired results. The hammer shown here is a swinging tool. The progression and exaggeration of our results are dictated by hand hammer weights, the height and ferocity of our swings, the amount of heat in the steel to be forged, the weight of the anvil beneath and the accuracy with which we can hit the same or similar desired locations consistently.

Picture 2
The near side off horn blow with convex face rounding hammer. Note the correct angle of our hammer handle at time of impact. The swing as well as gravity forces help drive through the steel to develop a forging technique.

Picture 3
Near side off horn blow with resting steel beneath. When hot steel is placed on the anvil, the blow is pursued as shown. This position is common for inside radius work, horseshoes as a toe bend starter, and many other turnings and rounding. From this I believe comes the name rounding hammer.

Picture 4
An offside (far) horn blow with steel resting. The convex rounding hammer face is being used to prepare an outside radius. After the bend is started, there is less chance of leaving hammer marks, so we can then switch to the hammer flat face working surface.

Picture 5
Atop the horn with rounding hammer. Now we are forging as opposed to bending. We can create shoulders in the steel, add length or width while narrowing down the forging. The hand hammer and anvil beneath are working solidly together. Find that sweet spot . . .

Picture 6
Near edge half face flat. Good for setting down single shoulders.

Picture 7
Near edge heel. Excellent for single or double shouldering.

Picture 8
Near edge half face angled.
Inside radius
Clean up blows to exit a curve.
Helpful in aggressive non-vertical shouldering.

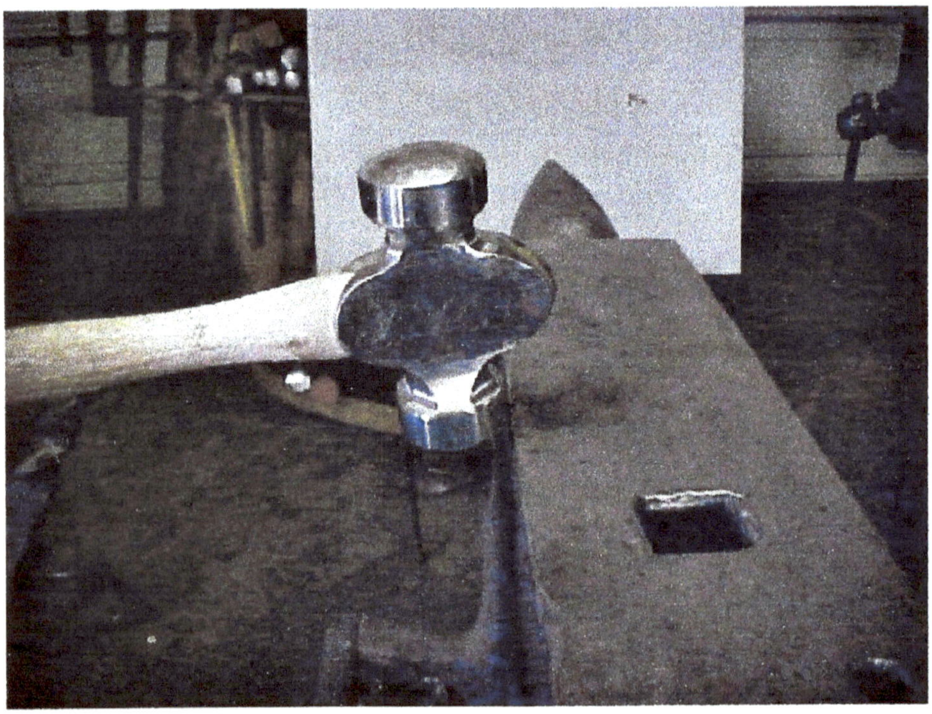

Picture 9
Off of near edge hammer toe.
Good for inside radius bending and finishing.

Picture 10
Full face flat. Good for leveling, straightening, general forging.

Picture 11
Far edge half face flat. Great for shoulder set down, clean up and innumerable possibilities.

Picture12
To far edge hammer toe flat.
Excellent for truing a thinned section after single or double shouldering.

Picture 13
Off of far edge heel blow. Good for shearing, offsetting, pinching, bending and many other options needing hefty or light impact comfortably.

Picture 14
Aside step, full face flat. Great for forming, short 90 degree angle starts.

Picture 15
Atop step, half face flat.
Set down portions requiring an angle other than perpendicular to the billets length are more easily adjusted here.
Anvil heft provides greater recoil.
Our eyes can view the developing angle more readily.

Picture 16
Along step angled. Achieving a "V" cut in square, round or hexagonal is possible here by ourselves.

Picture 17
Atop step angled half face. This blow works very well when finishing a single shoulder set down or bottom "V" cut.

Picture 18
Off of step heel blow flat.
Very helpful for pinching and truing up single or double shouldered material when accuracy counts.

Picture 19
Off of step with cross pein. The
amount of distance between working pein surface and step shoulder beneath at impact will create many variations.

Picture 20
Atop horn flat face of hammer blow. Great for trueing, forming, dressing, light drawing, bend correction, etc. This blow is used when an inside radius must be cleaned out while keeping a flat outside radius. The anvil horn is the inside while the flat of the hammer is your outside.

Picture 21
Atop horn rounding face of hammer blow. Used for heavier drawing and trueing up material. When the smith drifts the swing and impact near or far in slight increments, minute to dramatic result changes will occur. This also applies to straightening tools and implements.
The point of hammer impact to the work and position above the horn is all telling.

Picture 22
The near horn flat face toe blow.
The hammer toe is the focus here. Mainly for adjustments and light duty forging. The hammer toe typically has less effective force and accuracy. Impact from swing inertia and gravity are also reduced to improve accuracy. Miss hit blows will be felt through our elbows; so careful use here is advised. Our view is obstructed as impact occurs so arms should know and feel where the hammer toe will hit without seeing all with our eyes. To be used lightly yet with specific purpose.

Picture 23
Shown here is the far horn flat face heel blow. Great for turning many shapes. A two to three degree rotation of hammer blow at the point of impact to work makes this a turning hammer. Convex round or flat hammer surface is used for this. The hammer heel is a power house of inertia in hand forging. The swing should not exceed all security of the steel beneath. Regulate the swing carefully. Stock dimensions, heat requirements and billet securing requirements are all important here.

Picture 24

The rounding (convex hammer working surface) hammer shown is used over a hardy hole. This is good for upsetting and bending. The hot steel beneath and hammer impact control will determine your results.

These and many other hand hammer blow combinations will be discovered as work continues. Practice your observation and remembering skills for what each does with a given dimension of steel (thickness and/or width). How much and how little an effort of hammer blow is required to achieve the desired result of each swing and impact. Which position prepares the work for your successive procedure. Observe which blows are handier to achieve a shape; remembering that we have many ways to find a result for our work, but only some make the best sense.

THE RHYTHM OF WORK

Many times a smith may need to attempt many different types of projects to approach an efficient rhythm for their forging: Tending their steel in the fire. Retrieving a piece at its optimum heat for malleability. Utilizing that heat on the anvil or die to maximize its properties for shaping. Next, realizing when the best heat has passed for the immediate project segment and should go back in the fire.

This type of eye for judgment can easily be considered a skill in itself. Some may choose to keep hammering well after the ideal heat has passed. The best idea has been to consider having two or more similar forgings going at once. While one is under hammer, the other is waiting in the fire over moderate heat.

When single run or one time forging projects are in progress, some limitations exist. As heat is diminished in the steel, a break is taken to reheat it. The greatest advantage to this is but a chance to collect our thoughts and plan ahead while tending the fire. Let efficiency become a habit rather than an anomaly! This can be a great asset and a skill to be honed as our career advances and evolves.

Developing a reliable swing with our hand hammer is essential. Can we hit the same place on our forging project and hit it consistently? As the steel is changing shape, can we still pick a spot and hit it directly? This should occur with purpose and conviction. Regularly finishing one segment or setting up the forging for the next. Many times we should have prepared well ahead of time to set up our current forging step to prepare for the successive. All completed under a hand hammer. Remember that one hand is the power (the hammer side) and the other does the forging (the tong side). The tong hand holding your hot/cold steel lets the power come to life. Now enabling it to operate with all the fervor gravity will allow. Tong manipulation will dance a tune when the hammer swing is as a dependable machine.

For the timing of heats and balance of effort to complete a project there is a rhythm of work efficiency which can be achieved. Many smiths do not realize that a feel for their process at hand must be looked for! Each should acknowledge the timing of their heats. A balance of fire time for any given forged unit itself. Take fire or forge welding for example. How much heat is the forge giving you for the size billet placed in it? How is the fire doing? Clean or dirty with clinkers. . . how much forging needs to be completed in the next heat? All these factors combined should be considered to find your work rhythm. Once these factors are identified, they can be honed to perfection. Figuring out where the weak link exists and cognizantly helping that factor to fall in line for efficiency of our work day. These factors refer to the pace of our effort. Gauging all sequential progress to flow logically.

WRITTEN MUSIC AND HAND FORGING

WRITTEN MUSIC AND HAND FORGING

In the relationship of a hammer blow sequence to an orchestra's written music composition many similarities exist. In a sonata, minuet or full concert movement each has delegated notes for each bar (between measures) for each instrument. Mine has been the violin, which has no frets on the finger board. For the notes to sound proper they must be listened to for sequence and intonation. As is the anvil, each musical movement has a sequence and pitch (sound) requirement before the end result has public appeal and acceptance as a complete orchestral composition. As our forging project continues there are segments when the big hammer comes in for heavy blow, much as a tympany drum to create thunder and make for radical change in a passage. Many segments often require near silence for a solo effort. Some written notes are sticatto (short and sweet) or pictatto (crisp). Maybe we need slow and melodically smooth (legato). For music, all is written for continuity of progression, but what of the blacksmith? Most are best formulated by a series of blows in sequence of anvil position by a hammer of familiar heft. Any particular smith can develop his/her own written sonata for a forging. Much as notes of music, we listen and watch the progress. This is what makes things repeatable and reliable.

Even a master chef has some favorite recipes that when altered but slightly don't taste quite the same.

For your forging sequence of a tong, set things up with your favorite rounding hammer of appropriate heft to swing comfortably. Pick a section of steel much longer and slightly larger in cross section that required to complete your tong. Long enough to be hand held after heating a segment. Heat a short section and take it to the anvil. With your hammer, map out the blows to complete the forging <u>setup</u>. Not for the entire project, but to create enough symmetry and isolation of parts of the tong so that everything is allocated in mass.

The position of steel on the anvil and placement of blows can be written as follows:

	Anvil	Hammer
1.	Near edge	Half face 2 blows
	left quarter turn of steel	
2.	Far edge	Half face 3 blows
	left quarter turn of steel	

Moving up the steel (away from the smith)

3.	Far edge	Half face 4 blows

We have now set up a tong jaw rivet boss and shoulder for rein-emergence.

Now we are writing forging music to a great extent. We even get to keep the end result longer than a passing moment.

CUISINE AND THE BLACKSMITH

CUISINE AND BLACKSMITH

For the master chef there may or may not exist a written recipe for each dish. Many meals are produced from an informal combination of ingredients. In creating their desired texture, color and flavor constitution, each corner is turned as required to reach a savory end product. When and at what temperature are all considered as each ingredient is added? Do we incessantly add delectables or aim our efforts with a specific end product meal in mind?

For the blacksmith, profound understanding and acceptance of how much impetus can be imparted consistently from the swing of any given hand hammer. How much mass of forged material can be moved with the swing of that hammer. Should the billet be smaller? Should the hammer be larger? Is the billet heated enough? Was it too hot for the position it was held upon our anvil? Was the right style of hammer face employed (small or wide faced, flat or radiused)? At what point is it time to switch to a different hammer. These and more are all factors which develop the smith's recipe for outcome success.

Here our wheels are in motion. The coal is burning, the steel is cut. We have an end product in mind. We are now using our best judgment, swinging a familiar hammer, using known anvil edges and faces on occasion. Next switching out to new hammers and anvil surfaces to look for tidy results. Some call it "a dash of this and a little bit of that"!

TONG WORLD

TONG WORLD

A pair of tongs in practical use are an instrument used to keep its operator at a safe distance. Their structural shape and weights are as varied as the operators that put them to task. The length and heft each tong assumes encompasses a number of important aspects. They include:

- The mass of the material it must carry.
- How much heat will radiate from each billet (longer reins may be required)
- Type of fire used to heat material:
  Coal: A deep firepot in a coal forge may require a narrower bit/jaw to facilitate manipulation for thorough heating. Slipping through coal/coke chunks like a porpoise in calm water.
  Gas: Material selected to make a tong from should respond well to continuous reduction/oxidation type flame, i.e., little or no slagging over from radiant billet heat, or furnace insertion and retrieval; slagging over will at some point affect the tong ability to grasp securely.
- How much manipulation is required of the material they hold, heavier hand held tong to simply pass project from fire to anvil/swedge block/power hammer die block. Lighter tongs that will be turned, repositioned, opened and closed many times during each project heat.
  Super heavy foundry tongs cradled by overhead block and tackle chain will require proportionate upgrading. Reins are often held together by lock rings to secure the billets instead of the operator's grasp.
- Personal preference; some smiths are most comfortable with a short rein and a wide grasping placement. Their forging position is almost on top of their work. Others will complete their forging best when reins are relatively long and remain but ½" apart when attached to their intended work and blacksmith.
- How much welding will be done with the tong? Often a longer rein is an asset. The radiant heat and flux spatter is thereby farther from the smith. This alone can vastly improve forging confidence/hence end results.
- Strength/skill/coordination of the smith.
  Length of reins as a counter balance for some tongs will feel as though someone ground four or five ounces off of them simply by where your forging hand holds them when supporting any given billet/project.
- Do your tongs feel balanced when empty?
  Then awkward when grasping? Quite often hand position closer to or further from the rivet is not the answer. Many a tong's jaw simply lack enough mass in the jaws to help the counter balance aspect, whether empty or full (holding material).

- Will your tongs be used for one project or a continual part of your daily operation? Effort to find proper mass in front of and behind the rivet does as much for tongs to provide comfort and accuracy during use as any other aspect; right up there with how securely they hold your work.

- One time use tongs are entitled to no less effort in craftsmanship. Their downfall being the burden of helping a project simply to get closer to completion; not disposable, just at the mercy of form needing to follow function. The every day tongs should feel as an extension of the smith's appendage. As a hammer is in one hand, the tongs fitting comfortably/confidently in the other. As the tongs are being used, why not put an extra 10% of our concentration abilities on the tong hand for two or three heats?
- For balance sake: Ask ourselves, are the reins too heavy? Are the jaws too light when holding work? Start by looking at all the rein material behind the rivet to see where, if at all, they can be lightened. We don't want to compromise integrity/strength of the tool, yet if some material should go, let it come from the rein thickness first rather than width. Head for the bench grinder/belt sander and spend some time thinning them down a little at a time. Use the tong a couple heats; notice any differences and continue to lighten or stop when desired results occur. (tongs with less weight disparity in front or behind the rivet when holding your work are close to ideal in production/repetitious applications.) Hand position, fore and aft, especially on the reins will be a paramount consideration based on our body make up, eye sight, etc.
- Building balanced tongs. This can be a tough proposition since all tongs are not built while holding their eventual load. The mass and hence volume of steel selected to build the tong blanks from should have ample heft for their intended tasks. This is entirely up to the individual smith, yet remember, extra mass can be as much a hindrance as not enough.
- Play with the volume of steel in your tongs to determine how much is needed to support a given mass of hot metal to be forged. The jaws should make it through a solid heat holding their load without becoming hot and showing signs of fatigue. It is normal for tong jaws to heat up during use. For me, it is not acceptable to watch a jaw move or bend, or the reins sink closer together when in use.
  Put strength where it can do the most good, i.e., rivet boss, shoulder offsets, rein length. A well constructed tool means just that.

A TONG FOR ALL SEASONS

A TONG FOR ALL SEASONS

Tongs are an essential implement used mainly to keep its user at a safe distance for manipulation of their work. Whether material being handled is hot or cold, thick or relatively thin, short or conversely long, the purpose of the tongs is universal. To aid in control and hence passage to a finished product.

When possible, I opt to cool one end of a steel project in water so that it can be hand held during forging or bending. Yet to retrieve material from the forge (gas or coal), again tongs are needed for the partial quenching before it can be hand held. A very wet water soaked rag can be used for this if no other tongs are available; as if the very first pair of tongs are being forged for your shop.

During many forgings the shape of the project may change thereby making it necessary to change tongs. For safety reasons, as well as forging efficiency, much consideration should be given to this phenomenon. Especially for production work where the same path in forging will be trailed repeatedly. When the tong is not fitting securely, stop at that crossroad. Build the correct tongs for that segment of your forging. Modifying an old pair of tongs no longer used for a different project also qualifies.

As a surgeon enters a good operating room there are many standard instruments needed. A complement often considered to be full of essentials, all counted and coalescent for order of usage.

Conversely, the smith often learns to do without. To make do with the minimum. Tools strewn hither and yon, many not sharp or loose handled. Take a moment before jumping behind the anvil to access the tongs and tools situation even before steel is put in the fire. Ask ourselves, will there be a big bump in the road for a given forging?

Picture 25

Pictured is a complete storyboard of our flat jaw tong making steps. Each piece was made from the same dimension of steel, 5-1/2" of 1" diameter hexagonal (jack hammer bit steel). The steel was normalized, no hardening or heat treatment.

Each step includes only one additional phase in forging from the previous. The pattern runs from top left and down, then to the middle and down. Our finished tong pair is to the right. Making a consecutive step map of your favorite projects will solidify the knowledge for future reference. Story boards of projects which are not forged on a regular basis will not be so easily forgotten. The order and magnitude of the steps can more easily be examined for best logical progression.

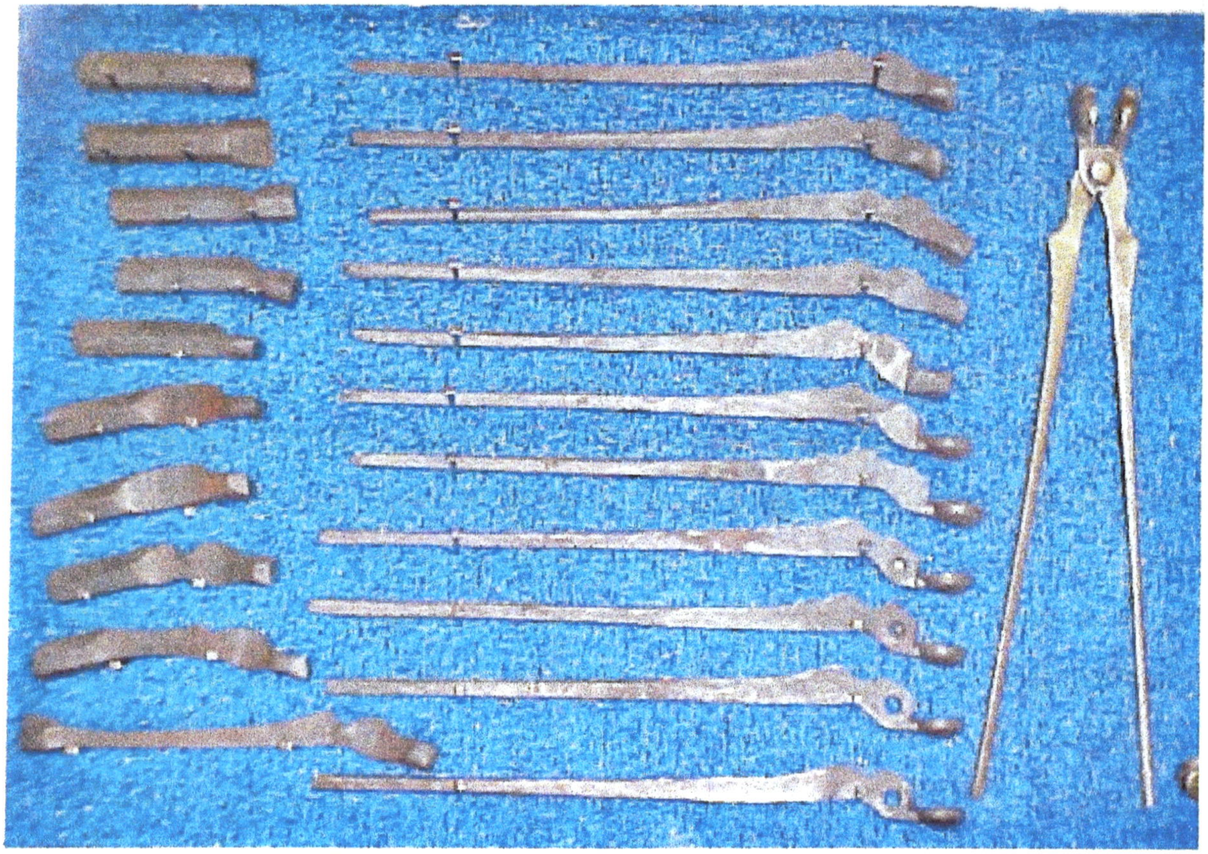

HEAT TREATING

HEAT TREATING

This is one aspect of the blacksmiths work that cannot receive enough attention. How often is a beautiful tool forged only to have a short lived career? It may have all the aesthetics, weight and size proportions that were intended. Probably ground and polished to virtual perfection, yet the integral/internal constituents have been ignored. These constituents are those that provide impact resistance, wear resistance, longevity, and safety for the operator.

We as smiths are not metallurgists in the truest sense, but we must touch upon their prowess and education to find satisfactory results in our tools, tooling, fabrication, forgings, and many times thought process. This thought process refers to how materials are handled, what they can and cannot do. How some steel is brought up to correct forging heats. What temperature is appropriate to begin and to end forging. How must a tool steel be cooled prior to heat treatment. How much heat can a billet of a specific tool steel absorb before deterioration results? Loss of carbon or other elements. Malleability limits and/or limitations. All factors which can and should be investigated when uncertainty or "critical" applications dictate necessity for accuracy. Referral to steel manufacturer recommendations and instructions may be needed. Also suggested use and application documents. Many have done the research. We just need to find and use it correctly.

With an appropriate heat treatment to an identifiable type of steel, we can impart a wide variety of toughness. With regard to knives, a blade can be hardened for sharpness of edge, and then tempered back to provide inherent malleability characteristics. A knife that can be sharpened for a razor edge, yet used as a pry bar when needed, to flex then return to its original straightness. Many large hunting knives are hammered through a bone or joint cartilage of a carcass without breaking or suffering irreparable ill effects. When the appropriate degree of hardness is achieved the blade can then be tempered.

New steel purchased from a mill or distribution warehouse usually comes in the form of annealed material unless otherwise specified.

The grain structure of the steel is paramount. For these factors to work correctly and reliably most metallurgists will agree that on a microscopic level, steel will harden best for the blacksmiths purposes when it has been annealed after forging is completed. Annealing is a slow cooling (descending) heat from a specific temperature over a specific period of time (duration).

I shall cover the metallurgical grain structure aspects only in brief, not for lack of importance, but because much of this cannot be seen with the naked eye. Its relevance to a tool's durability is essential, yet I have chosen to focus on elements with which we can tangibly create a noticeable cause and effect by practical means for the smith's purposes. Protons and neutrons, relating on an atomic level, are recognized throughout much of the smith's heat of forging and later cooling. The goal is to train our eyes, ears and hands to recognize when and where the appropriate heat presents itself. Where it can be forged and when is the appropriate time to stop will then become apparent.

These guidelines are presented to help the smith become aware of some of the cues dictated by the steel before them. Many times we may have heard "while not looking, an answer appeared." My preference is to be alert. Mentally recording what the heat is doing. How much heat in terms of degree and duration is required for a given thickness to forge it by hand or under the trip hammer? Material purchased as tool steel (medium to high carbon content) can be expensive so why ruin it...

Using this information, the heat imparted to the billet for annealing is raised to critical temperature and slowly lowered. Most are done from a billet color ranging from cherry red to orange; annealing is then achieved slowly and surely. When all forging is completed (the last forging heat) it is a good idea not to overheat the billet and anneal. Annealing is the slow cooling process of the billet used to help reorganize the grain structure of the steel. A great deal of stress relief is also possible after intricate forgings are made. This greatly reduces the chance of fracturing during heat treatment hardening. Done primarily on carbon or materials known as tool steels.

A fine grain in the steel reoccurs when cooling from a specific heat over a long period of time. This duration of heat soaking time should be long enough to provide uniform heat throughout. Not too hot or too cold, just uniform in heat. The duration of this cooling process for annealing can range from several hours to several days. The chief variables are.

1. The type of steel forged (carbon content, alloy formulation, etc.
2. The thickness or thinness (external/internal dimensions)
3. Overall length

When steel material such as knife blades are to be annealed after forging, the thinness is quite the factor determining how the steel is cooled. Often we choose to anneal in wood ashes or lime. A large heat sink 2-3 lb. piece of hot remnant steel is placed in the anneal can first.

The steel 30 gallon can is filled with the preferred or available medium; the can also has a steel lid. The lid placed on the can after the steel is buried in the medium is used to inhibit air drafts. Drafts occurring during annealing can and do reduce uniformity in the steel grain structure and hence tool quality after hardening.

Thin items being annealed have very little mass to hold heat. Many, on their own, cannot absorb and retain enough heat from the fire to cool in the medium very slowly. A number of solutions for this have been found.

A large chunk of scrap steel is placed in the forge to gather heat before the annealing stage is initiated for the forged billet. The chunk is heated to the correct temperate as the forging dictates. It is extracted from the fire and placed in the bottom of the wood ash or lime can first. Remembering that heat rises, this will provide a residual heat to all else place in the wood ash can above that chunk.

Another means of annealing is to also leave the billet in the forge after reaching a good cherry soaking heat. This works well in a coal forge. For production work it is better in a gas/propane forge. The front door is closed and very little draft results. An electric furnace will also work well for slow cooling. Residual heat from the forge liner/coal will slow the cooling process quite nicely.

Anneal can mediums include:

1. Wood ashes
2. Nursery grade lime
3. Vermiculite
4. Loosely-packed fine clean soil

Quite often we may have heard mention of finding a critical heat. In the old days, this may have had several different definitions.

1. To watch the color of heat in the steel change.
2. To know what temperature has been attained.
3. To use a known temperature stick that vaporizes or changes color at specific heat levels.
4. To use my all time favorite; a magnet.

For hardening

1. Air
2. Water
3. Salt brine bath
3. Oil hardened steel

Virtually all of our tool steel is hardened. Annealing is done for stress relief and grain reorganizing after forging is complete. For impact resistance in swinging tools, add to that good wear and hot working characteristics in many set tools. Virtually all of our steel is hardened out of a good annealing stage. Mild steel and low carbon steel may not fall under these guidelines.

Molecular structure and grain development will not be covered in depth for that is a subject into itself. It can and should be investigated and studied in depth, if desired. I have chosen to delve into aspects which are most practical for the smith's conventional shop procedures. That which can simply be directed and effected by our eyes, ears and hands to produce a <u>safe</u> tool for use under blacksmithing or deemed simple engineering applications. All done from a coal forge or gas-powered furnace without a thermocouple (temperature gauge). Again, out of the annealed stage.

We will need a good (strong gauss) magnet to be our eyes for this segment. My favorite is a cattle stomach magnet held in a pair of one-half inch (1/2") hollow-bit bolt tongs.

With any ferrous metal that is cold, it will have attraction to the magnet. As the steel is slowly heated, the molecular activity increases and hence heating progress is checked regularly by touching the magnet to it.

As the steel is leaving its natural black/grey color and beginning to show color from heat, it will be noticed that less magnetic attraction is exhibited. It's telling the smith that the molecular structure is changing. Because a good annealing is behind us, we know that a fine grain structure within the steel exists. As the magnetic pull lessens, the grain structure is becoming stirred up. Martensite to austenite often termed metallurgical, and all points between. Molecules begin to move, then grain structure changes. This is all done on a rising heat.

Our ideal point to quench the material is when the steel first shows no signs of magnetic attraction when touched by the magnet. Some steel may need an immediate quench; others may require an extra second or two in the fire for an added 50-100 degrees. Practice with your steel to find an ideal.

Regardless of the quench solution (air, water, salt brine or oil,) the hardening medium should be selected with the manufactures recommendations in mind. For unidentified pieces of steel, some experimenting will need to take place. Using relatively small pieces will minimize waste for this experimentation.

The steel is quenched out to near cold and can therefore be hand held. A brief moment is usually taken here to test the steel we just hardened. With a moderately used 10-12" mill file, the steel is held in hand on a work bench. The file is stroked flatwise across the hardened steel to test for hardness. For our swinging or set tools, the file should skate across their striking surfaces with moderate pressure.

The hardened surface resists filing and therefore, has accepted its heat treatment. For somewhere even close to good impact resistance, the tool should repel the mill file test.

After hardening, the tool should be tempered.

Items left in the hardened only state may begin to show surface cracks. This can occur within minutes or may take days or weeks to happen. Deep cracks may develop if the quench medium was too cold and cooled the steel extra rapidly. Our quenches are made to be room temperature of 60o F (Fahrenheit) regardless of the yearly season. On colder days, a large remnant of steel is heated in the fire to act as a heat sink. It is then plunged into the quench medium before heat treatment is commenced for that day. Ice chunks in the quench medium are <u>not good</u> for this type of heat treatment. If oil quench is being used, care should be taken not to warm your heat sink stock too much, as surface flashing or fire can result. Having a lid for the oil quench or simply plunging the heat sink all the way under the oil surface when warming up the quenches are helpful. Using an oil that has flash resistant properties is also a possibility.

The tool has been file tested to show <u>no</u> soft spots on its surface and is ready for tempering. Working surfaces are polished to chrome to view color progress as heating occurs. The removal of all rough grinding scratches is not essential at this point. The subject of immediate attention is to get your tool started in its tempering heat. Remove just enough scale and loose slag to see your colors run on the clean steel portions. Some steel requires tempering at a heat above or below the range at which colors will show brilliantly on the surface. Check this before any heat for tempering is delivered.

Many air hardening tool steels may require heat of tempering to a temperature at, above or below the temperature from which they will be used. For example, a hot/cold chisel being built. Tempering is not only done for the type of steel being forged, but for its intended use! The type of steel may be identified for each of these tools, but its form (shape) and function (operating temperature) mandates your appropriate heat treatment. This is what contributes to greater safety and longevity for its intended use.

GOOD THOUGHTS

I will try not to cover too many specific projects for heat treatment as far as the tooling built is concerned.

Many other skills and background needed to accomplish specific forging tasks will become apparent. The smith will develop a better understanding for his/her steel and its application. Some trial and error may result, but with vigilant attention and recognition, better decisions can be made. The basis for those decisions has already begun. ☺ Keeping a small note pad handy in the shop to record temperature, time duration, color of steel, ambient air temperature, date, and steel dimensions will all help when we get a tool that came out good. We are now developing a road map to make the journey successful and repeatable. We are also learning more about how our own personal forge works.

Learning to access tool use and working backwards from there has been a tremendous asset. Before any steel is selected or cut we ask ourselves questions; is the goal a set/swinging tool? Does heat resistance take precedence over impact resistance? Are both essential? If we are to forge a hot working tool, the sharpness of working edges can be as important as wear resistance. The type of steel and heat treatment process implemented has much to do with our end results. Heat resistance and long term sharpness (durability) are directly connected to your heat treatment, therefore annealing, hardening and tempering are all encompassed.

TEMPERING

For tempering hammer heads, old worn out drifts that fit well into the handle holes will suffice. I prefer to heat tools with holes, through the holes. This assures that as the heat of tempering reaching the surface of that item has been thoroughly soaked throughout its mass. Tools or items without a through hole are started at a very low temperature in a preheated forge or furnace. Our forge is occasionally turned off as the items are placed inside for tempering. The residual heat of the refractory (forge liner) provides us ample rising heat to temper. As the colors run to our desires, the tools under construction are extracted. Mention is made that this process is for air, water and salt brine hardened tool steel.

After tempering, when the item is extracted for cooling, after water and salt brine hardening, they are plunged into a tepid water tub and remain until cold.

For tempering oil hardened steel, the process is slightly different. The type of oil preferred is regular used crank case motor oil. This includes from 10-30 wt to 20-50 wt. No engine cleaners or detergent agents were added to the engine before it was drained. This plain oil is preferred because it will flash for us to determine the proper degree of heat for tempering.

There are many heat treating oils now available for this which have flash inhibitors. Somewhat self inhibiting for our method, yet they have a special purpose in other applications.

After an oil hardened piece of steel has been quenched in the oil to harden, it is removed and allowed to drip. These drippings fall back into the oil quench tub. While dripping, the gas forge is turned on. A coal forge works fine for this as well. As the dripping slows, we grasp it with the tongs and present it to the fire. Good ventilation in the shop should be ever present. The tool has been hardened and now we are tempering with oil as our indicator. The flashing of the oil will look more as a dance than our normal forge fire. Rotating the tool to allow all areas to receive equal opportunity to receive heat. You will know when the tool's heat is uniform for the oil will not flash up. Its own dance of fire has stopped. The second that no more flashing is occurring, the steel is removed from the fire, examined for a brief moment visually to view any wet oil spots. If all looks fine, it is placed in a warm water tub to stop the tempering process or allowed to simply air cool. This last cooling selection is mainly dependent upon the mass of steel. Thin knife blades have much surface area and very little internal mass, so allowing them to air cool is usually a good selection. Cooling them faster from a tempering stage to water bath may produce surface cracks. Conversely, a 10 lb. sledge hammer of oil hardened steel has a reasonable amount of external surface area and much internal mass to hold heat of tempering. The tempering may continue to soften the steel after extraction from the final heat. Because of the large amount of internal mass holding the heat, it may need to be cooled off quicker to stop the tempering. This can include many different means from placing the steel in front of a forced air fan blast to dropping it into a running cold water tank.

Remember that these suggestions are from the "old school", tried and tested, that have and do work They allow for great results under a low technology budget. They are affordable by most of the population in a blacksmith shop whether hobbyist or solid professional smith; quality results can be expected with some practice.

HANDLED TOOLS: SWINGING AND SET

There are innumerable numbers of handled tools which we use daily in our shop. Broken into two (2) categories, they include swinging tools and set tools.

Swinging Tools

Hammers, cross peins, offset peins, rounding, sledges, axes, mattox, rock picks.

Set Tools

Flatters, hot and cold chisels, hole punchers, fullers, creasers, hacks and side sets.
Before the forge is lit and the steel is selected, the smith should take a moment to make a few decisions. Ask ourselves a few questions:

- The type of tool that will be forged – swinging or set tool.
- The method of handle attachment – eye for wooden handle; eye for mild steel handle; wrap around mild steel handle; welded on mild steel handle.
- The mass or heft of steel required to make the desired tool. Do we have a piece forged or small enough to complete the head of the project? A larger project may require a larger diameter or longer handle to keep the smith at a safe distance or provide a safe/secure grip in hand.
- Are all these materials currently in the shop? Take an inventory check. Look through the bone yard to pull out several pieces of the most appropriate handles and head steel. Brief materials inventory check can save many disappointments. This will tell you if the project can commence and be finished.
- Narrowing the field – do we have the tools needed to make the next tool or project at hand? Example: It makes the best sense to have extra punches large enough or small enough for good through eyes, in hammer heads or in handled punch forgings.

Eye punch picture 26

Different Handles All For the Same Purpose
Punches

Type	Bottom View	Handle Material	Union	Side View
Eye Punch		half round mild Steel 5/8"	welded 312 stainless DC reverse Arc	
Eye Drift		coil spring 5160 round drawn, necked down	riveted joinery strong shoulder	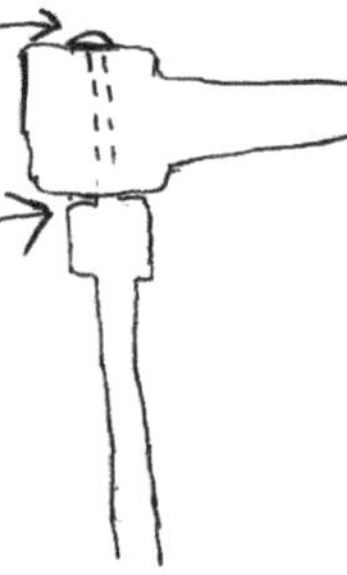
Farriers horse shoe fore punch		1/4" x 7/8" mild steel Bumped up welded end	welded 312 (5/32) stainless DC reverse polarity Arc 110 Amps	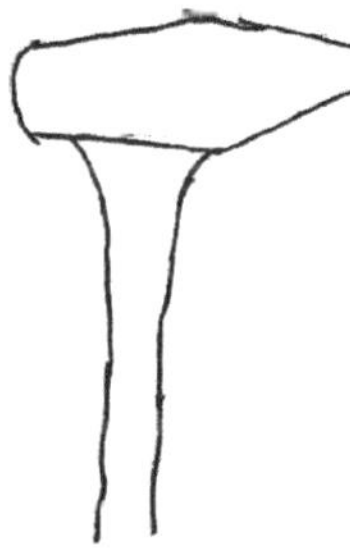
Tong rivet hole Punch		1/4" x 7/8" mild steel Draw in throat to 1/4" x 3/8"	welded 312 stainless	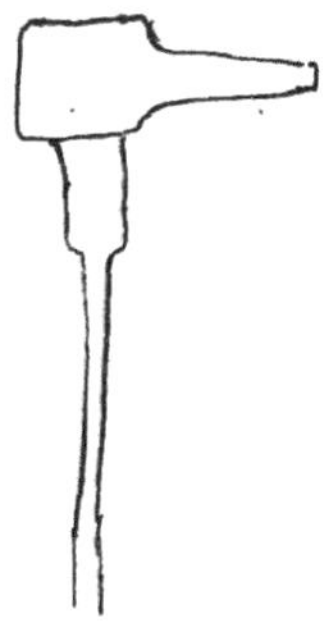

The Type of Handle Materials Available

Wooden – hickory, fruit, osage orange, and eucalyptus. All 1-1/2" to 2" straight branch stems rasped down to eye and hand size.

What is the Right or Correctly Built Tool?

Our definition – one which provides for safe use time after time, no chipping or marring, or excessive mushrooming under its intended use. Remember it takes more time to make repairs than to do it right the first time!

Steel

Straight and long enough to be a handle. For our purposes, all steel handles are only for set tools, not to be swung, just for being struck. The shock transfer to our hand, wrist and elbow dictates that when the set tool is struck it may sting. To help resolve this, an oversized piece of steel will need to be thinned down in the same neck portion to dissipate this transfer of energy. Only mild steel is used for handle materials (such as A36) with low carbon. This is appropriate not only for cost savings, but also to inhibit the sting and stock transfer to our hands. Miss hits on the tool head are more forgiving when this exists, also the handle can bend and save the smith much trauma! Wrapped around head handles of 1/4", 5/16", and 3/8" round stock have proven good in service. This 8-14 inches approaching the hand is folded under to be doubled for mass and security in hand.

Many of our steel handles are welded to heads for accuracy in use; most are 1/4" x 3/4", 1/4" x 7/8", and 1/4" x 1". No eye punching is implemented; merely a working tool with a handle welded to it after heat treatment. Some are then retempered for stress relief of the welded section. Our preference is 312 stainless arc welding on reverse polarity; DC negative. Excellent for attaching mild steel to tool steel. Remember to wear good protective clothing; gloves, welding glasses, helmet, and as dark a light shielding lense as can be found!

Riveted Handles

Steel handles with an eye hole through the head of set tools works well also. Not everyone may have a good arc or mig welder, so we may need other options.

Picture 27

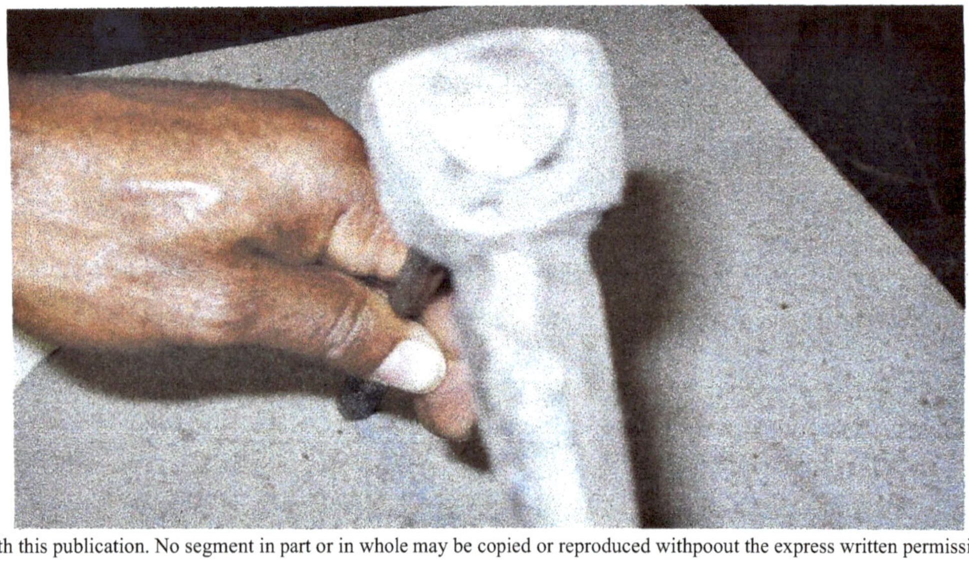

For these, a small oval or round hole is punched through the tool head; 1/4" up to 5/8" diameter depending on the type of tool. Our preference is round quite often after the initial hole is punched. An oversized punch with a blunt point is inserted into the side of the eye where the handle enters the tool. The exit side of the eye is resting on the anvil or bottom die block. Imagine trying to stuff a football into a thimble. This gives you the appropriate picture only slightly exaggerated. This punch is driven into the eye to part and smooth the steel handle entrance way.

On the handle eye exit side, a large jackhammer bit is driven 1/8" inward. This will leave a cross hatched depression deep on the sides of the eye hole. These depressions can also be put in with a handled hot chisel. The furthermost point or the nearest point of the chisel is inserted in the eye hole. The chisel is struck and then again moved around the hole to create the same cross hatched pattern. These marks are deepened as required by the tool head size and hence mass.

Preparing a Handle for The Tool

Round stock is ideal when hand forging a handle for this type of head. When trip hammer, steam or air hammers are available, many other steel handle materials can be easily modified however.

The main consideration is to have a section of the handle up near the eye entrance which is too large to go through the eye. This forms a shoulder to wedge the head on the handle and prevent sliding toward the smiths hand when in use. The portion passing through the eye is necked down to just fit nicely. Now we're into joinery in the truest sense.

The rule has been to pass one and one-half times the thickness of your handle material beyond the exit hole. The thickness of the material passing through the hole being the base measurement. If the diameter of the handle material fitting in the hole is 1/2", add 1/4" to this. So 3/4" should be protruding beyond the exit hole of the tool head after insertion.

Picture 28 - Shouldered Handle to Go in Eye

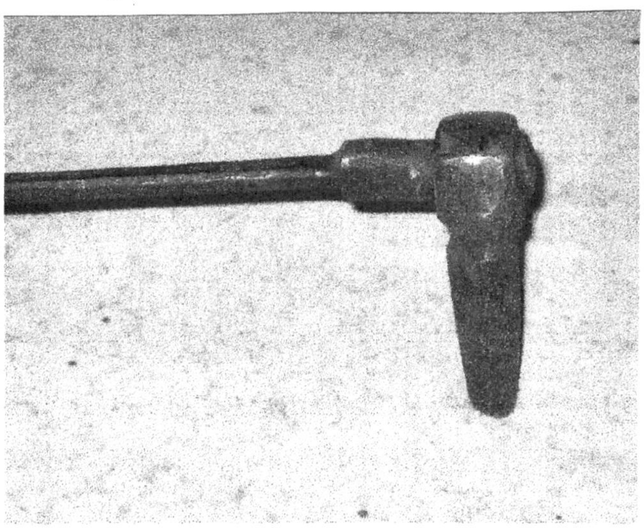

Occasionally when the entrance hole is not perfectly seating the handle, we can correct this. Merely heat the shoulder section and hand forge the imperfections to enable seating. The intended tool head can also be tapped on the handle to find and mark the troublesome spot. Draw down another shoulder facing the hand held portion of the handle. This will also double as a shock relief point for comfort when the tool is in use. Leave this section thinner than all portions of its length.

The hand held portion is quenched in water to become cold. The portion which will be inserted through the tool eye is again heated to deep red color. The handle is then hung – hot end pointing upright in the box leg vise and the vise is tightened. The large shoulder preventing the handle from going all the way through. The tool eye is now suspending the handle from the vise jaws. The set tool head is now placed over the red hot section. The head is driven onto the handle with two or three accurate hammer blows. Often these blows will need to be on both the working side and then to striking side of the tool to keep the head centered on its handle.

Next Page Top
Picture 29 - Shouldered Handle in Vise to be Put On

When many handles are to be attached in this way using a short section of mild steel pipe or our version of cup tool is recommended. The blows can be delivered harder and more accurately. Our cup tool is 4" tall with a welded on mild steel handle. The depression in the middle of the cup allows the eye steel to pass inside while seating the tool under construction. This tool is also good for adjusting tong rivets when they are being set. When tools such as the cup are needed, it's best to stop here and make one, as it will have many uses as your forging progresses.

Next Page Bottom
Picture 30 - Cup Tool to Drive Head Onto Handle

Back to the red hot steel handle, drop the head on while the handle is supported by the vise and handle shoulder. Drive on tight, free hand or with the cup tool. The protruding segment is upset into the exit hole of the tool being made. Being sure to push the section with your hammer blow at first, then later the rounding side of a rounding hammer to spread out the upset portion to fill the cross hatched depressions.

This is what keeps the handle from swiveling while in use – locks it right in. Once in a while the head may tend to bounce while the protruded section is being upset. So have a pair of tongs handy which will hold the head being set and prohibit the bounce factor while upsetting the exit lock handle material.

Picture 29

Picture 30

WELDED ON STEEL HANDLES

For each smith, there will be some preference with regard to length, width, thickness and hence heft. One large consideration should be given to appropriateness. Lightweight handles for lightweight tools. Search for a balance in the feel of the tool handle which enables good finesse in movement and manipulation of your tools. A handle which lets you keep connected with the depth a tool is sinking to when being used; as with punches, helps the smith know when and where it has reached the bottom of a hole. Not only visually by his hand hammer blows not moving it progressively; but by our sense of feel through the handle. Under a power hammer, our keenness and senses do much better when the feel through the handle can be relied upon. As for eyes and ears, they must now do more when working with a power hammer and open dies. The handle must enable the smith to turn up their connection with the work at hand. An appropriate handle can and does help with this aspect. But successively, he/she can monitor their progress by the feel through the tool's handle. This can and does require repetition to find; however, it does profoundly exist. Quite an asset that is sought after once the smith learns what to look for. By adjusting each handle to accommodate this feel, a good continum in work results can follow. Each of us has a favorite tool for each project. What makes one implement better than another when all else is equal?

For many punches that are welded to steel heads, the options are limitless with regard to size and heft. A chief requirement in size is to limit heft of tools to be driven by hand hammer or sledge. The larger they are, the greater the inertia requirement to move them. Remember to leave some section near the head in the handle thinned for miss hit blows. Accommodation in this handle to twist or bend without causing serious injury to the smith or rendering the tool unusable. Sections through which the tool travels that have many irregularities in both width and thickness are attempted much more favorably when this tool is in use. Mild steel when thinned can provide a true safety net to bend considerably cold before breaking. We must just search for what is the appropriate degree of thinness in this handle section. That cross section which allows a bend when needed, yet secure and sound implement usage.

Picture 31 - Thinned Steel Handle for Shock Absorber

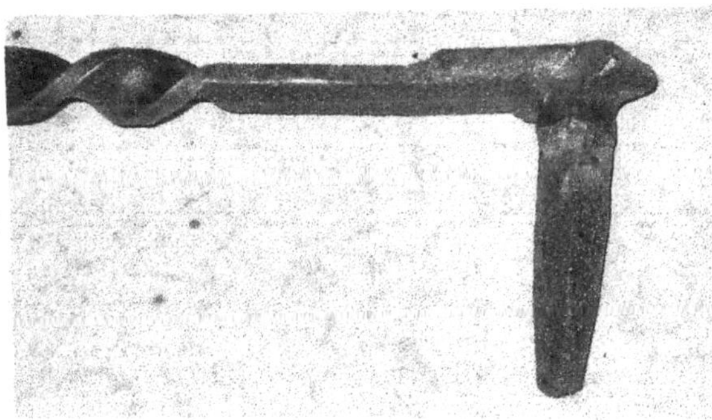

The size 1/4" x 7/8" is a favorite of ours for hand punches of many purposes. I am told that mine are larger than most; this dimension has served very well for many a student and professional smith. (Photo: Welded Handle Punch 1/4" x 7/8")

Handle Prep

The portion to be welded to the tool head is heated, quenched out to within an inch and a half of this end and upset. The extra mass here helps to provide a slightly greater cross section for welding.

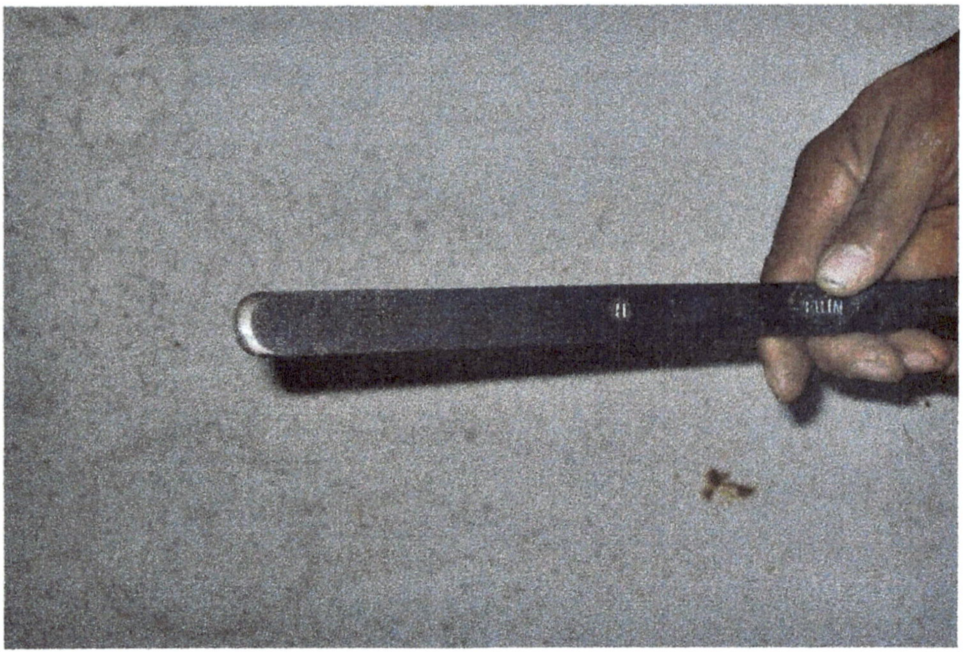

Picture 32 - Handle Ready For Upsetting Weld End

Under hand held conditions, the tools are welded to handles providing the 1/4" thickness in vertical position. For power hammer applications, the 1/4" thickness dimension has done best in horizontal position. There are many reasons for this. A few include:

Power hammer vertical height preservation – more space for the dies to rise and fall above the forging material. More of the tool is available for actual use without handle obstruction.

Less sting - to our hands in general forging from shock transfer.

Ease of punch extraction - should a tool become lodged in a billet during hole punching, more surface area is available to ease removal in many cases.

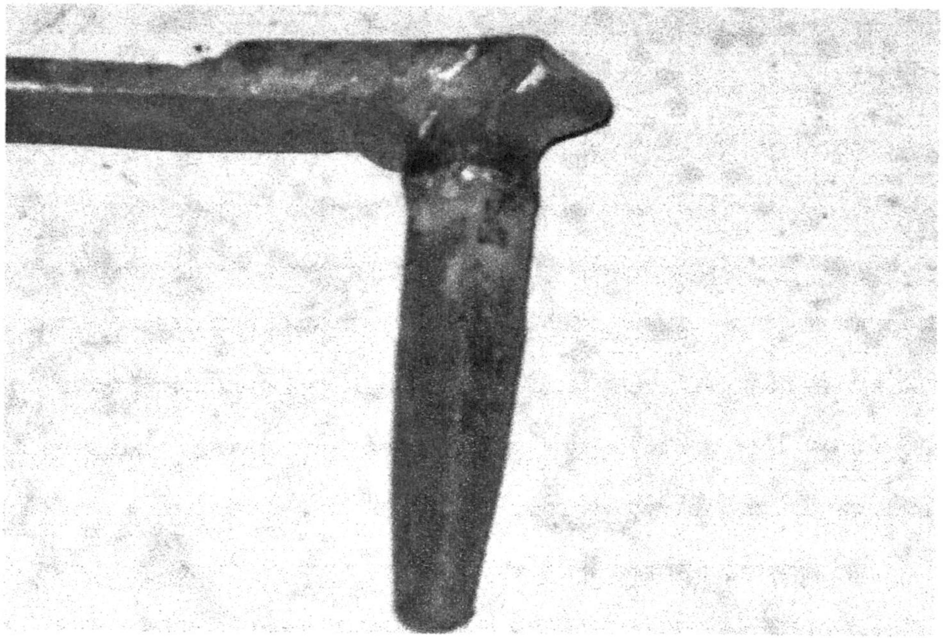

Picture 33 - vertical and horizontal 1/4" thickness

Safety – this cross section positioned to the realm where it is needed to bend most easily.

Punch extraction – from deep holes (2" to 3") often the horizontal positioning of the 1/4" x 7/8" handle aides in the removal. Remember the handle cross section is attached to be wider than it is tall. The chief drawback to this positioning of steel handles is that your line of sight is slightly inhibited when accurate work in specific areas are needed under power hammers. A very minor drawback when the benefits are considered. Because the handle has a good feel to it, we can circumvent this relatively easily. Nothing should be inherently built into our handle design to cause disconnection with the work being forged.

Accurate work often requires a tight handle fit to its head. Many smiths may become accustomed to the amount of movement that a handle dispels because of previous use. It then provides negligible difference in end result when used. Loose handles usually require multiple strikes to help a tool find its mark. This typically requires much more work than should be needed over the course of a day. Loose handles like some fine wines have an acquired taste, but when we find a good one that works, one usually sticks with it….

Many times a tool such as a flatter is moved many times to level imperfections in a work piece. We can probably get by with improper handle fit and looseness but imagine one which lets you know what a section of steel is doing without looking at it with an eagle's presence of mind. We probably would prefer tools that will allow some reliability through feel.

Considering the amount of use the tool will receive over its career, we believe even single run or specific use tools should receive good care in forging and quality handle attachment for best and accurate usage. Repetitive use tools must have excellent handles and the best steel available for their heads.

HANDLE COMFORT

At the end which is hand held, much can be done to provide comfort in the set tool. Some individuals choose to get a handle on and go to work. The old tough guy method dictates, "it's on there so let's use it." A favorite has been to find a means of comfort without losing the tool's accuracy feel to the work. With the same 1/4" x 7/8" dimension handle steel already attached to the tool head, we are ready. The handle can range from 14" to 26" in length, depending upon the tool it is attached to. The protruding corners closest to the smith are ground off smooth and rounded.

A good quality automotive heater hose section is purchased. It helps to bring a piece of the handle material to the auto parts outlet to find the appropriate size. Depending upon the brand, pick one which will just go on an inch or two dry before becoming impossible to push on any further. That's the one we want to use. The length of hose each tool will require is variable for the smith as well as the tool. A 20" handle of 1/4" x 7/8" has done well here with a 13" hose.

Next page.
Pictures 34 and 35 - Punch With and Without Hose on Handle

For round stock handles, they can be folded under to become double and again driven through the heater hose.

The Hose Handle Grip

To begin, grind sharp corners off the hand held end. Push the handle into the hose as far as possible by hand. Squirt a small amount of lubricant onto the steel where it enters the hose, i.e., water, liquid soap, vegetable oil, penetrating oil, etc. The section of hose which has the steel entering is now wrapped with a fold of leather; just one square foot is enough leather. The leather wrapped sections are now put between the jaws of our box leg vise.

With the working end of the tool now pointing upward, the vise is cranked down snugly, but not a death grip. The leather is there to provide friction to the auto hose and protect it from being torn in the vise jaws.

With a brass hammer now drive the working end of the tool through the hose an inch at a time. As the handle goes through the hose, stop. Open the vise and reposition the leather wrap between the vise jaws as needed. Re-lubricate once in a while. Close the vise onto the handle and continue hammering on the tool head until the handle finds the hose end. Wipe excess lubrication from the hose and we now call it ready to use.

Next page.
Pictures 36 and 37 - Hose onto handle

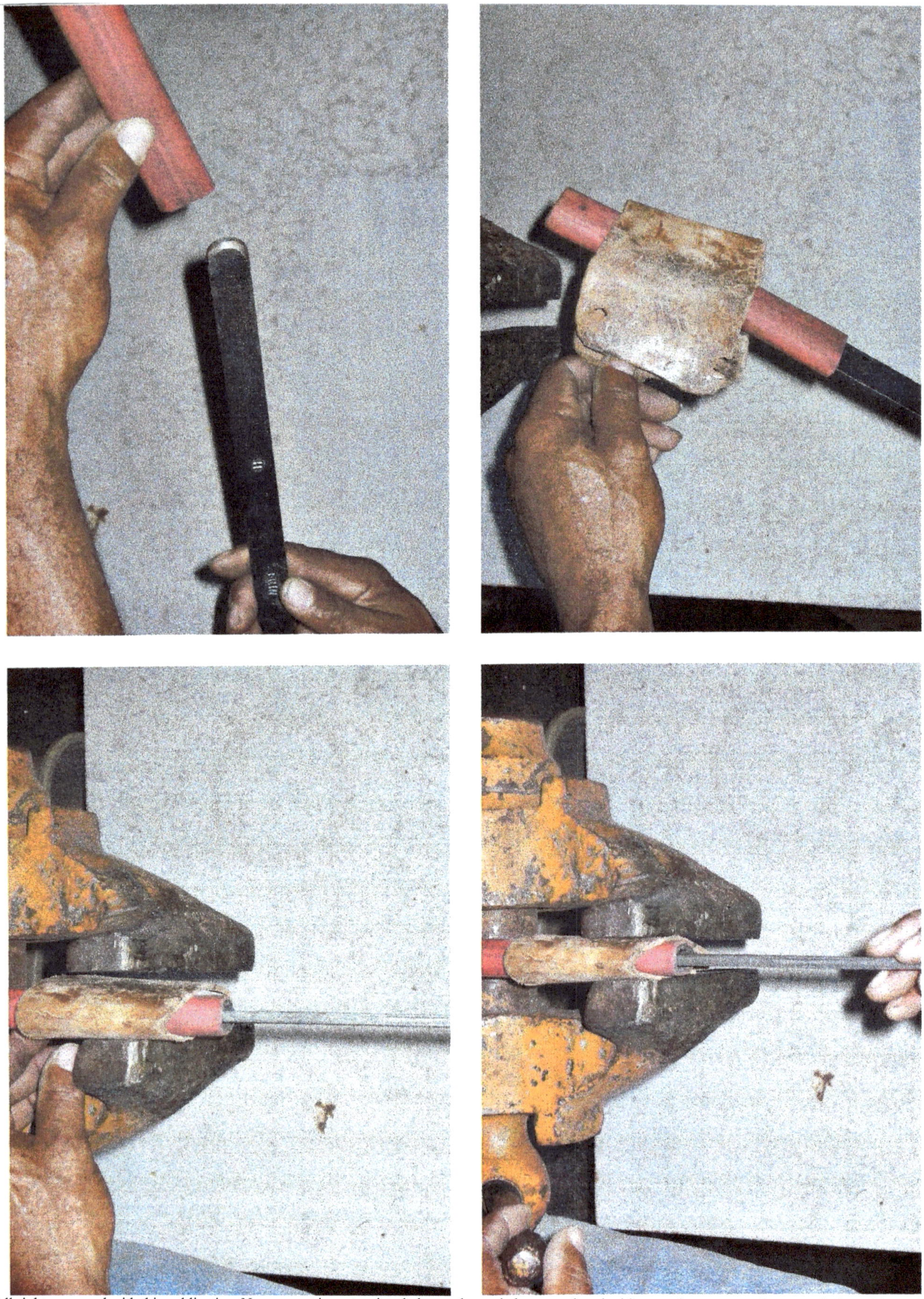

HAMMER HEAD
HEAT TREATING PREFACE

If the steel has enough carbon in it, we can impart a wide variety of hardness or toughness in the tool. Other elements in the steel affect its metallurgical properties, chromium, molybdenum, etc. These constituents dictate the temperature it can be forged from, also, the elasticity characteristics for forging.

Because we have practiced much forging, it has become a habit to look for the steel to move so much for a given blow of the hammer. In drawing or shaping that steel, if it resists moving at the same forging heats, we know it to have more carbon content and choose a heat treat accordingly. Much unidentified steel runs through our shop so attention to its forging helps determine a heat treat process for that individual piece of steel.

For each chunk of steel, there are usually a few right ways of hardening and tempering and many wrong ways. The durability of a tool has plenty to do with the heat treat process adopted; assuming a good piece of steel was chosen to build the forging from. The heat treating process is one large area that drew me to making our own tools for the shop. When a blacksmith has to purchase the same tools over and over, something must be wrong. Is it user error? Most likely, yeah, but this is left up to the user to prove first. After forging all pieces, big and small, they are placed in the lime or large wood ash can for annealing. Slow cooling in these mediums helps reorganize the grain structure and provide some stress relief, depending upon the type of forging. The billet can even be left in your forge or furnace to cool off slowly with the refractory or hot forge coals. This is not as reliable unless the fire is ignited once in a while or stoked with wood to burn out slowly, but better than nothing. The wood is only placed in a coal or coke forge. Not in the gas furnace.

Rough grinding is a good idea now. The steel is in a softer metallurgical condition and will not tear up the abrasives as much. Be careful not to over heat in spots. When heat colors such as brown, purple, or blue appear on clean polished steel it has been over heated. Keep the steel cooled off or stress fractures may show themselves in the spots where color from grinding occurred. Colors again range from straw to brown to blue and purple; all are undesirable in this stage. The steel should be kept cool enough that no colors show themselves. A periodic room temperature water quench has proven good for this purpose.

Picture 38

HEAT TREATING HAMMER HEADS

Picture 39

The ash can for annealing.
Steel 40 gallon with steel lid.

Picture 40

An offset pein hammer just out of the wood ash can and fully annealed. Most hammers will require a full day (24 hours) to cool for hand handling. They are placed deep in the ashes at a deep cherry heat after all forging is complete.

Picture 41

A group of cross pein hammers out of the annealed stage and given their initial grinding. This is the extent of their prep before our hardening stage.

Heat Clips – Because many of our hammers are forged to have thin sides and elongated eyes, there is some chance of fracturing during the hardening process. The clips shown are used to plug the eyes of tools ready for hardening in water or salt brine. Note the eye sides have not been ground to remove slag and will aid in protecting and slowing the hardening process. These clips will serve to further slow the hardening of this type of eye. Fracturing is rare to non-existent when these clips are used correctly. On tools with a heavy center section (3 to 4 pounds) of the eye, such implementation is usually not essential but also helpful. We make clips for this purpose of A36 mild steel. The heating and quick cooling of this steel is repetitious; though sacrificial they have held up well over time. They are forged to fit through the eyes and along their exterior as well. Each tool eye is plugged with these before being placed in the forge to pick up their hardening heat. The tool head is plunged into the quench with the clips still in the tool. For air hardening type tool steel, these clips have not been needed though could also be of value.

Picture 42
Heat Clips

Picture 43
A hand hammer made ready for our hardening process with clips installed.

Picture 44
Side view of **mild steel clips** installed through the hammer eye.

Picture 45
An assortment of mild steel clips for eye plugging. Keeping a variety of lengths, widths and thickness handy, no time is wasted installing them to fit all types of tools requiring heat treatment. The clips are used over and over again.

Picture 46
Side view of our **tool eye plugging clips.**

A cattle stomach magnet is placed in a hollow bit tong and kept ready. Finding the critical heat temperature to harden the hammers from can be a chore, and a color our eyes are not used to recognizing. When this temperature presents itself, the hammer will begin losing its zest for magnetism. Periodic checking with a good magnet will help find this appropriate heat and hence color. *See opposite page*

The tempering of tools achieved through **hot drift insertion** is of "Old School" technology. The results of which have proven excellent for each tool that is being heated thoroughly from the inside out. This process proceeds slowly and is very controllable.

Should one striking surface accept heat slightly quicker than another because of weight or configuration differences, there are solutions. Hold the hot drift in a pair of hollow bit tongs and present the heavier side of your tempering tool closer to the forge flame. Should one side of the tempering tool get way ahead of the other (one side showing straw and the other with no color at all), simply dip the side showing color in a quench bucket and continue the hot drift tempering process. *See opposite page*

Here our hammer has been hardened in water. The striking surfaces have been repolished to remove scale of heat treatment and make clean steel visible to view the temper colors. A clean surface to see the colors more clearly.

The temper colors are being drawn with a **red hot drift placed into the tool eye.** Normally with a two-pound (2lb.) hammer head, four to five red drifts placed in the eye will produce blue in the hammer middle and light straw on the striking surfaces. As color runs to black on each drift, it is removed and another red one is inserted. While one drift is in the tool eye, three or four others are being heated in the forge. Each successive drift is placed into the eye opposite of its predecessor. *See opposite page*

Picture 47
A cattle stomach magnet.

Picture 48
Hot drift insertion.

Picture 49
Red hot drift placed into the tool eye.

Picture 50
Hammers which are freshly hardened and tempered are pulled from the water quench tub and dried. The temper color again is checked for uniformity. Cracks on the working (striking) surfaces will show themselves with reasonable clarity after the temper is drawn. Darker color lines in unusual places show the tool has become suspect.

Picture 51
Finish grinding which requires the tool under construction holding still while using a hand held grinder to expose clean and slag free steel, sometimes the forgings hold still and the grinders are moved upon them. This is achieved as shown. The underside of this hammer's eye is resting against the vise jaws to prevent undesirable movement and rotation.

Picture 52
Hand held angle grinders come in many sizes and shapes. Shown here are two (2) regularly used in our shop for stripping and dressing slag covered tool steel. At left is a
9-inch heavy 36 grit abrasive wheel. The right is a small 4 1/2" flapper disc. The grit here is 50, and the backing plate is soft and flexible.

Picture 53
Shown is a deburring wheel installed in a bench grinder. The inside and outside cone type support caps have been reversed. The coned portion is now pointing inside of the deburring wheel and causes it to center on the grinder spindle. For grinder wheels which have larger center holes, this simple modification can make purchasing or building countless large spindle collars unnecessary.

Picture 54
As a deburring wheel wears, it can become hard to keep in service. The guards and tool rests on a bench grinder may prohibit safe use. This wheel is ready for replacement. The deburring wheel is used to remove scale after hardening to expose clean steel.

Picture 55
This is a muslin buffing wheel which has separated from its side threads. It's time for repair, servicing or purchasing a new one as this condition makes loading the rouge very difficult.

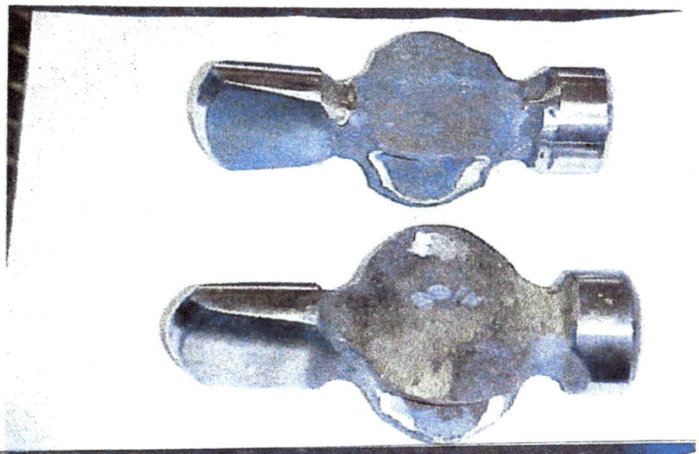

Picture 56

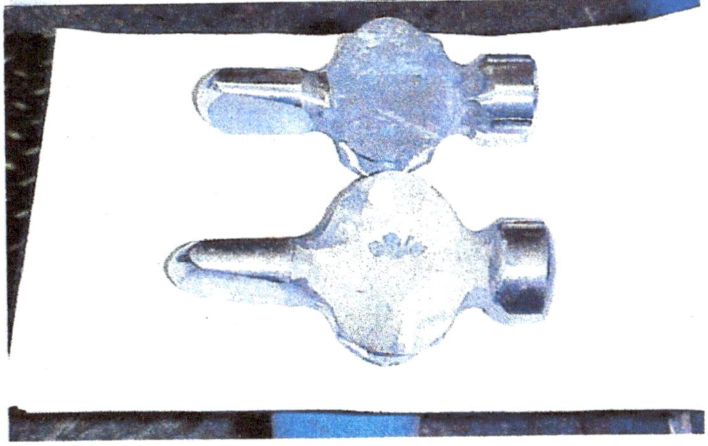

Picture 57

Two similar cross pein hammers at different stages of their heat treatment. The hammers above have only the striking faces ground and rough polished. The hammer will be hardened in water. Because the steel surrounding the eye is thin from the forging process, there is considerable risk of stress fracturing and hence, rendering it unusable. By leaving the slag developed from forging and annealing on this center section, satisfactory results have been common place. Just a thin protective barrier to slow the steel hardening process is all it requires. As with many of the forging steps, it's not what is taken or moved, but what is left and when. The eyes of the hammers are sometimes plugged with mild steel when a head smaller than 1-1/2 pounds is to be heat treated.

The hammer head below has been hardened and tempered. The slag covering clean steel surrounding the eye is now rough ground to remove. Should the smith desire to see all tempering colors run, the rough slag should be removed before the tempering process begins. Clean steel needs to be exposed to visually observe the colors running. Ideal is blue around the eye and straw to brown on the striking surfaces. This, however, is completely dependent upon the type of steel used and the hardening process (water, salt brine, oil, or air) implemented and the type of hammer desired (intended use). The temper colors just described make for a very lively and durable water hardened hammer for use in hot steel forging. A hot drift, one foot (1 ft.) long, is inserted into the tool eye to achieve such a temper color differential from the hammer middle and striking surfaces. Heating from the inside out provides a very complete and controllable tempering, for as the drift heat reaches the steel outer surface, all internal portions have already achieved the appropriate temperature. Both of these weighed in at 2-1/4 pounds when finish polished.

WHAT THERE IS AND WHAT THERE CAN BE

WHAT THERE IS AND WHAT THERE CAN BE

Quite often a friend or spouse may offer an opinion as to what they see–as in a realist's approach. Some may see what there is, others may see what there can be. We feel the smiths often fall somewhere in between.

The true innovators/inventors may tend to look at mass/volume current state or shape or even carbon content and then decide what it should be for the present need. This last thought process may consume a more scientific approach. Maybe even troublesome and take much of the fun out of our trade. But as experience evolves, this stitch in time may save more time; then becoming the sought after method.

We must all start somewhere in our development. Learning first to walk, then a trot and later to run. The preference, if you will is a nice brisk trot, for there lies efficiency. Allowing our mind and body sufficient opportunity to calculate and prepare while enabling progress. It also provides for reasonable recovery from a fall; e.g. miss hit steel. As long as we are born healthy, of sound mind and body, there is good opportunity for creativity.

Assuming there is purpose in our desires, part of the problem may be finding a way to make our dreams come true. Trying to bring an idea to reality has a different interpretation for each individual. As long as there are thoughts, clear or otherwise from our tradesman, there will occur varied results.

Once an idea has come to mind, it is often good to provide it with some support parameters. A means for that idea to grow roots; crawl into reality. To develop this idea, a picture or drawing is helpful. Sketch if you will, some or all of the elements. They can be partial or complete. As little or as much as can be rendered for tangible construction /fabrication. On many an occasion it has been conveyed that if you are able to draw an item you can forge that item.

Many smiths see this as a goal rather than a common pathway before project commencement. The purpose for placing thoughts in a drawing or picture is to help bring this picture into focus.

Many ideas can be brief or fleeting; possibly a simple moment of clarity passing. If we can find some means of recording that segment, all else may be built around it in time. Even if its purpose does not suit the immediate intention, the recollection may serve a past or future attempted forging.

Practicing a partial element of the forging can also be helpful. Quite often the student, regardless of skill level, will say "I can't see that," or "I didn't see that." Can't is not in our vocabulary.

The problem is usually not in visualizing the entire forging. It is the stumbling; the development of that forging which can be the problem. Try to find some solace in the steps you are familiar with. Try to see how they apply to the end product desired. Those one or two steps that need a bridge across them, over them, or under them, but hardly ever around them. Challenge those steps and punish them into simplicity. Try to isolate the elements which are familiar from those which are not. What makes the difficult ones a challenge? Is our drawing of this forging realistic? Are we just placing anything and everything on the anvil to attempt our forging?

Occasionally our steps are merely out of sequence, assuming none are missing. Some steps may need greater attention, both in detail and in movement of mass. Preparation out of one step making ready for the successive is randomly overlooked by some smiths. Keeping a focus on the end picture desired would be our virtue, not an altogether vendetta. Keeping focus on the altogether end picture should become and will be our virtue–adding experience and skill along the way. Learning or teaching ourselves to operate along these guidelines is useful for providing a traceable and repeatable pathway for product development. This is quite desirable.

WHAT THERE ISN'T

WHAT THERE ISN'T

What there isn't pertains to the number of hours in a day. Rarely can enough time be found to suit the smith. Very few of our projects can be taken from inception to completion in a single effort. This is to encompass one day as the parameter. This is to include what can be termed the Type A personality. Those which want things done yesterday. Frought with trial and error that would resolve if simply given a fair and reasonable trial period. Difficult to provide satisfactory results this way.

Our Type B person can provide for any and all the preparation required. To not waste a single piece of steel. Diligence and precision well before any physical project commencement. These individuals may think through all aspects of the effort. Take measurements before, during, and after all drawings are made.

Our particular preference lies somewhere in between these two thought processes. Finding one that works for us and enables productivity without undue frustration is now recognized. For many this may need some attention, acknowledgement, alteration and then incorporation. Bringing up our weaknesses both internally and externally after they are identified will improve any and all future forgings attempted. We each have a definitive portion of qualitative as well as quantitative effort within us. Rationing each asset wisely through due diligence is an art and skill unto itself.

DRAWING AND SKETCHING

DRAWING AND SKETCHING

I've heard it said many times that if we can draw something, we can forge it. As the riddle assumes, we have artistic drawing inclination proclivity and dexterity equal to our forging skills. Our drawings do not have to appear pristine in aesthetic presence. They are merely tools to help us unfold some of the layers of twists and turns needed to complete our project. What this serves to prove is that our minds eye can create and develop a clear picture of our vision which can then be replicated in steel or iron. Sometimes this will challenge or evoke a desire to reach enhanced proficiency in our forging abilities, but what if a question were asked: "Can I teach myself to forge in steel or non ferrous metal what can be drawn on paper?" "Can I learn to draw what can be forged on an anvil with these same hands?" Yes we can - for in drawing on paper, the element of eye/hand coordination has been removed. We are now all equal. When our sketches are teaching the hand where forged material should be hit with the hammer, or where held on the anvil a billet should be placed to achieve a specific bend; our eyes will develop a clear picture of where the steel should be hit to leave one step and make a bend to create the next step. All to get the smith closer to the desired picture. The more this is practiced, the closer our projects will resemble our drawings. We will learn to draw at or slightly beyond present forging limitations. The drawing of forgings which are segmented (many pieces forged then assembled) creates tremendous cost savings. Paper or chalk on a board is relatively inexpensive. Very little time and materials are invested, just brain power. Missing a step or ruining steel to be forged may get any of us behind a little. Three dimensional drawings or sketches are made in three views of the same step as needed. They include: End view; Side view;
Above/below

To start, try and focus your drawing for forging of one view. A freehand or ruler straight edge outline will work well. Even a combination of both is fine. Later, add as closely as can be approximated what would be the other views at each stage. Each view is carried through as far as can be pictured on paper or on the chalk board.

If exacting measurements are required, they are added lastly. A note or two for each step as to how you feel the step should be accomplished. Example: Near or far edge of anvil, when to mark steel, when to weld, bend over or across step of anvil or on horn. . . Cross pein or ball pein hammer in use. This segment of the drawing usually brings out some real creativity in us all. Trying to keep it all realistic may be a focus to contend with. But remember, there may not be any real answers – only the search.

If any segment troubles you through the forging, stop and examine the actual steps. Go to that portion in question. The forging and the execution of that forging may be fine, but the steps could be out of their best logical sequence. Helping one step prepare for its successor will become a feeling we learn to look for. When this continuity is missing, that too will have a feeling as well.

Forging a cross pein hammer and trying to put the wood handle through hole in last. Change this order immediately. The hole is punched as or nearer the first step. Examine the drawing you have made previously to decide whether one or two steps require some expansion. Often called an exploded view in some trades. If your drawing had two or three forging portions/steps combined into one view, expand it to expose even greater clarity. Turn over as many rocks as possible. Record what is working as well as what isn't. Give yourself a cross sectional view if needed to expose/record internal billet or material forging changes. The drawings presented in the text are as originally created by the author. They were developed for shop reference and soon became a small but vital library addition.

Hand Forging tong blanks without top Tools

Actual steel material cross sections to start

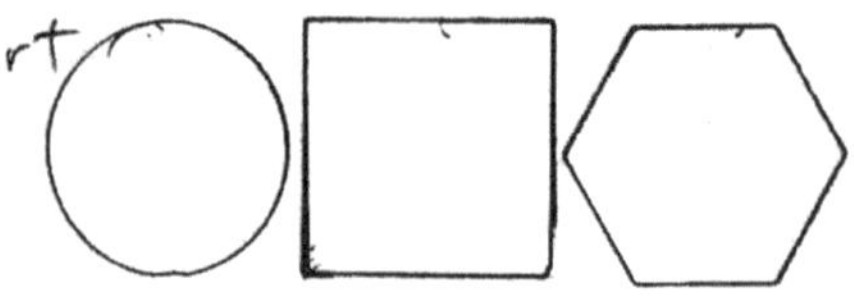

Materials: Good - Tractor implement coil, coil spring, 4140, sucker rod, titanium (VALOVR) low magnesium, Truck/car axle (unspiraled), jack hammer bits, rail road spikes

Progression:

Farriers Type: OK - Rebar, A36 (mild steel), wrought iron

Side view — Top view

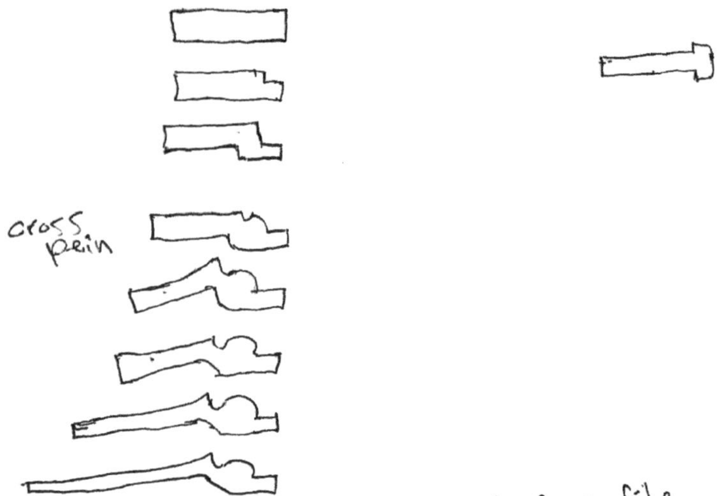

Rein Finish

Tear drop profile

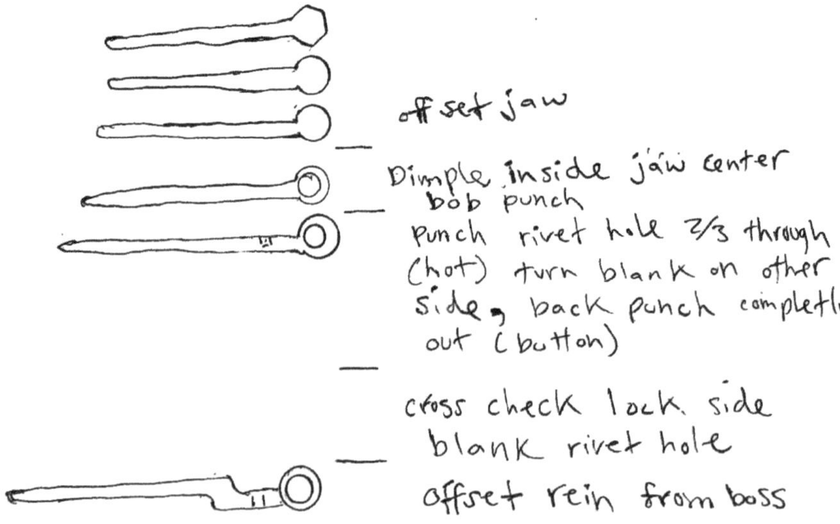

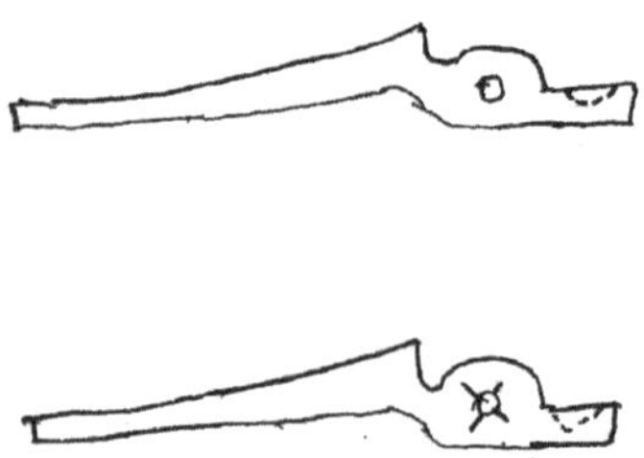

Hand forging Tong Blanks with Top Tools, set tools
striker with sledge power, steam/air, treadle.
hammers

Progression:

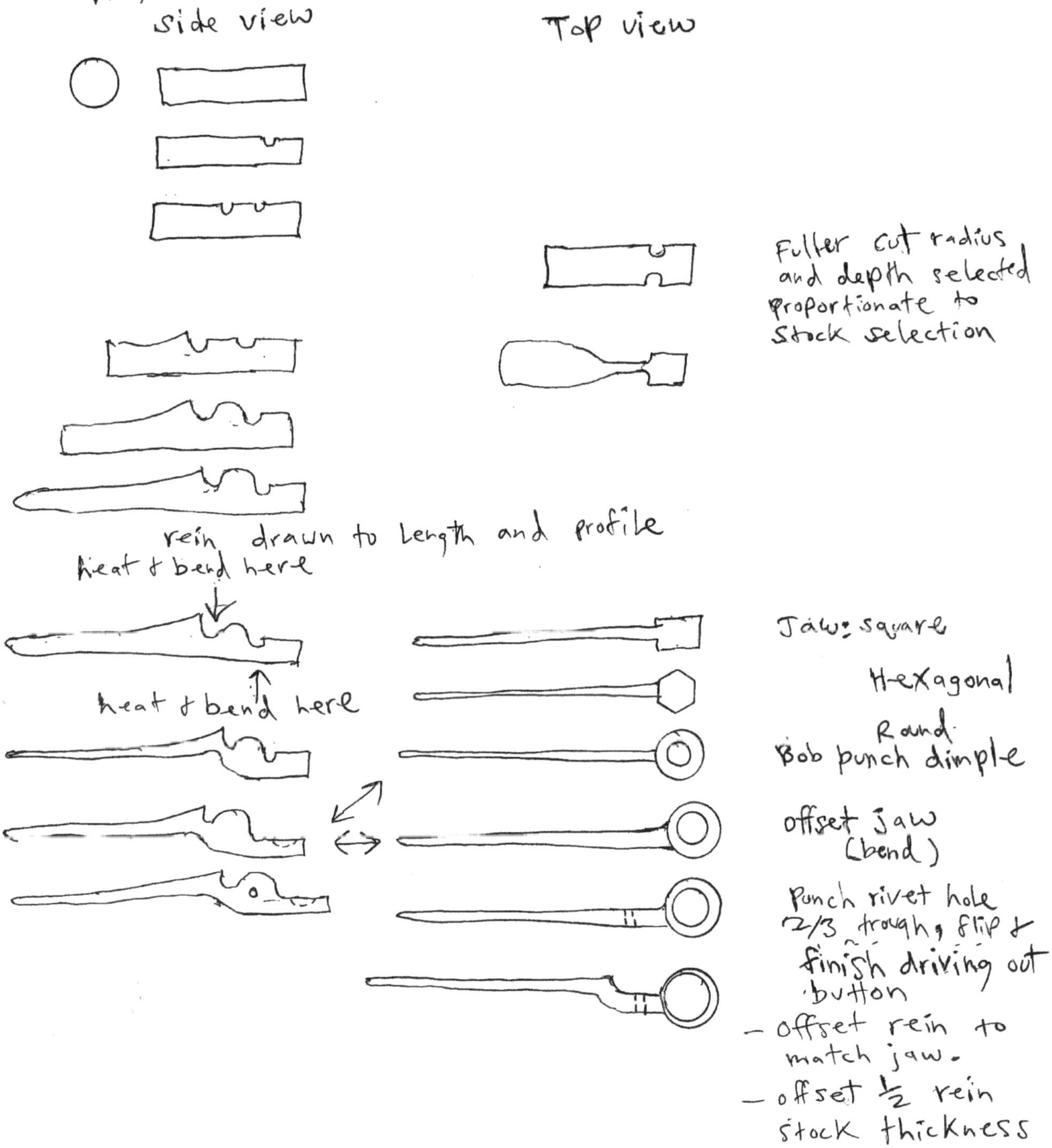

Swedging

By definition: Forging to change and displace volume to a desired profile. Overall mass per cubic inch is considered drawn only slightly.

By vision: Alteration of material external characteristics.

Correct starting material profiles are all varied but a most appropriate and efficient shape, volume and mass for each top or bottom swedging does exist. Search for what is reasonable, safe and gives sound results.

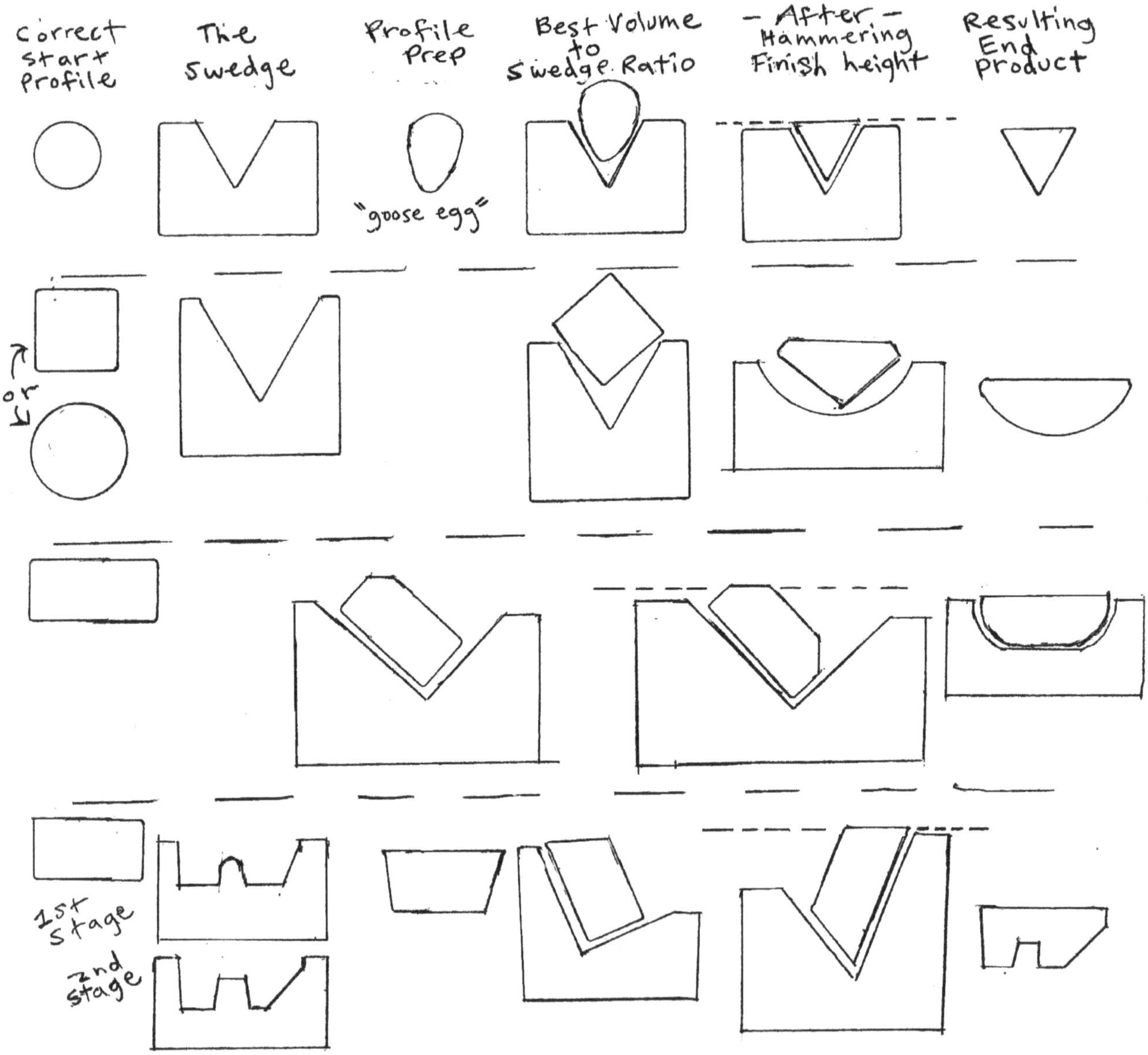

Swedging

Occasionaly top tool swedging is all that is required to achieve a desired contour or profile. Top tools are used before, during or after stock material is swedged through.

Hammer blows are rated and measured with our eyes and hands for the correct impact inertia to progress ratio.

Experiment with different starting steel dimensions to find an ideal. Which size is too big or too small to fill a swedge complete and uniformly. What mass is about right yet pleads for some pre-swedge prep.

Drawing → swedging a tong rein
sequence of end views:

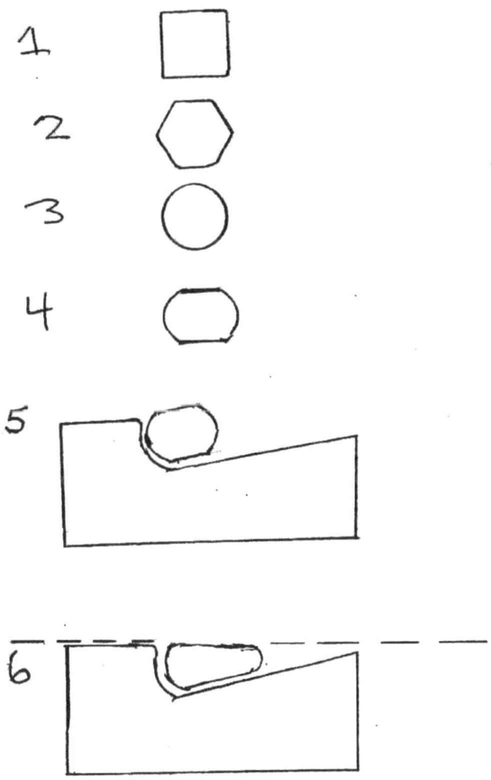

When a rein tapers in overall width and thickness from shoulder to tip the swedge will look as follows:

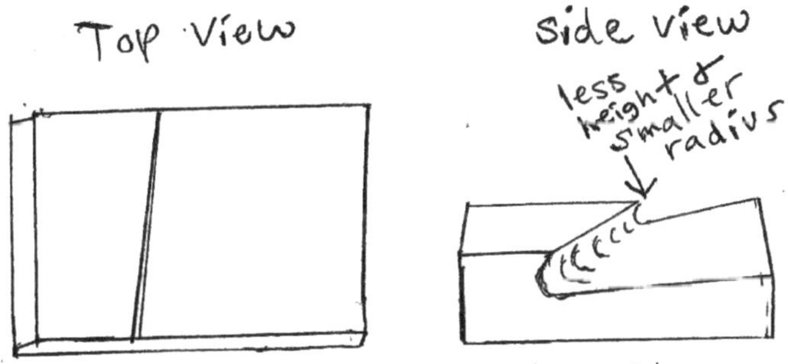

material is draw towards the smith as overlapping full face hammer blows are delivered to the forging.

"The Tear Drop"
a favorite rein profile pattern...

Special Bottom Die/swedge used To Produce Concave stock

Materials: Air hardening tool steel (eg. H-13, S7)

End view — Top view — Side view:

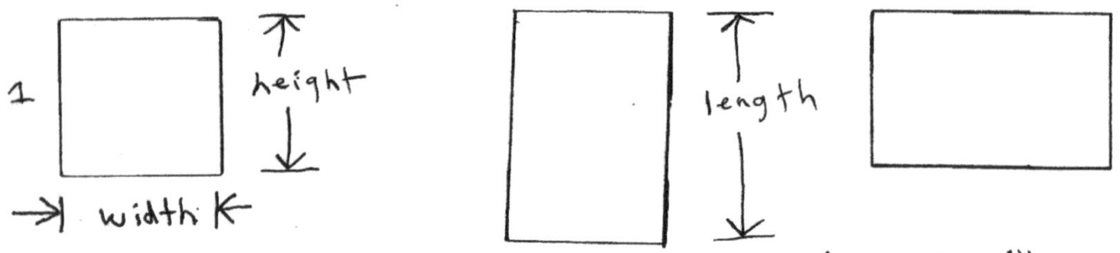

- The starting block selected which will become the swedge is both taller and narrower than desired for a finished product. Incorrect length to some degree is tolerable. Height will be reduced, length and width will grow
- Handled set tools such as narrow radius fullers, narrow set hammers and hot chisels are used. Many passes are made.

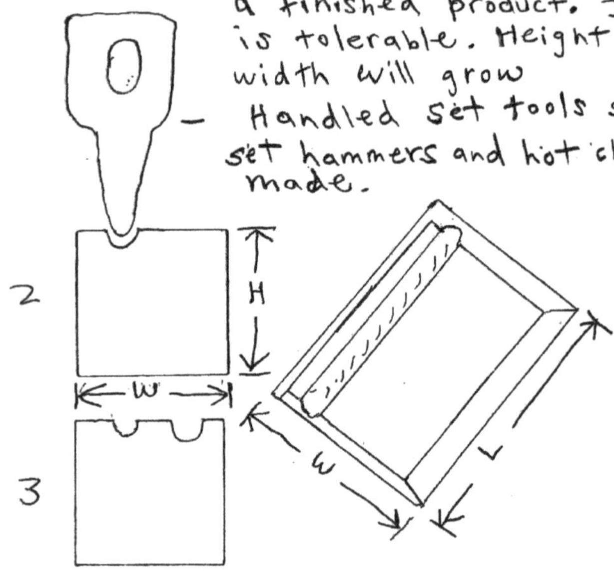

- Internal bottom depths are made to roll off lower at their ends.

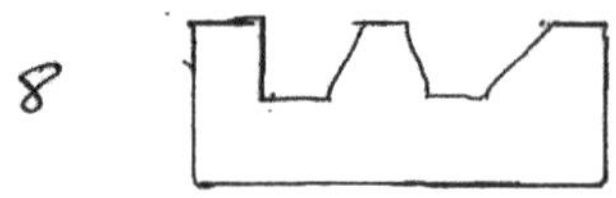

Swedge, Final Profile

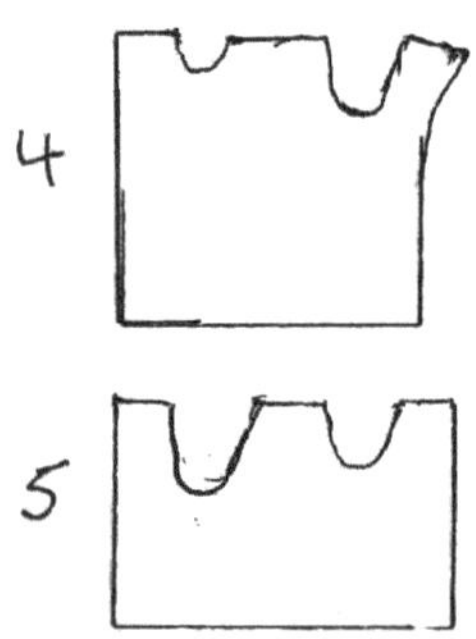

- As deformation occurs square the block back up and continue.
- Fullers which are offset in profile and crisper of radius are now run through.
- At this stage our depth dictates time to switch tools to finish.

Working ends are radiused longitudinaly to enable walking them any required forging distance.

- Anneal
- Rough dress
- Harden
- Debur
- Temper
- Attach tang (for hardy) saddle (for power hammer)
- Redress and test run a blank piece of orange hot steel through.
- Round or rectangular stock hammered and fed through with overlapping full face blows. multiple passes as needed.

Left to right

How Our Holes Are Punched

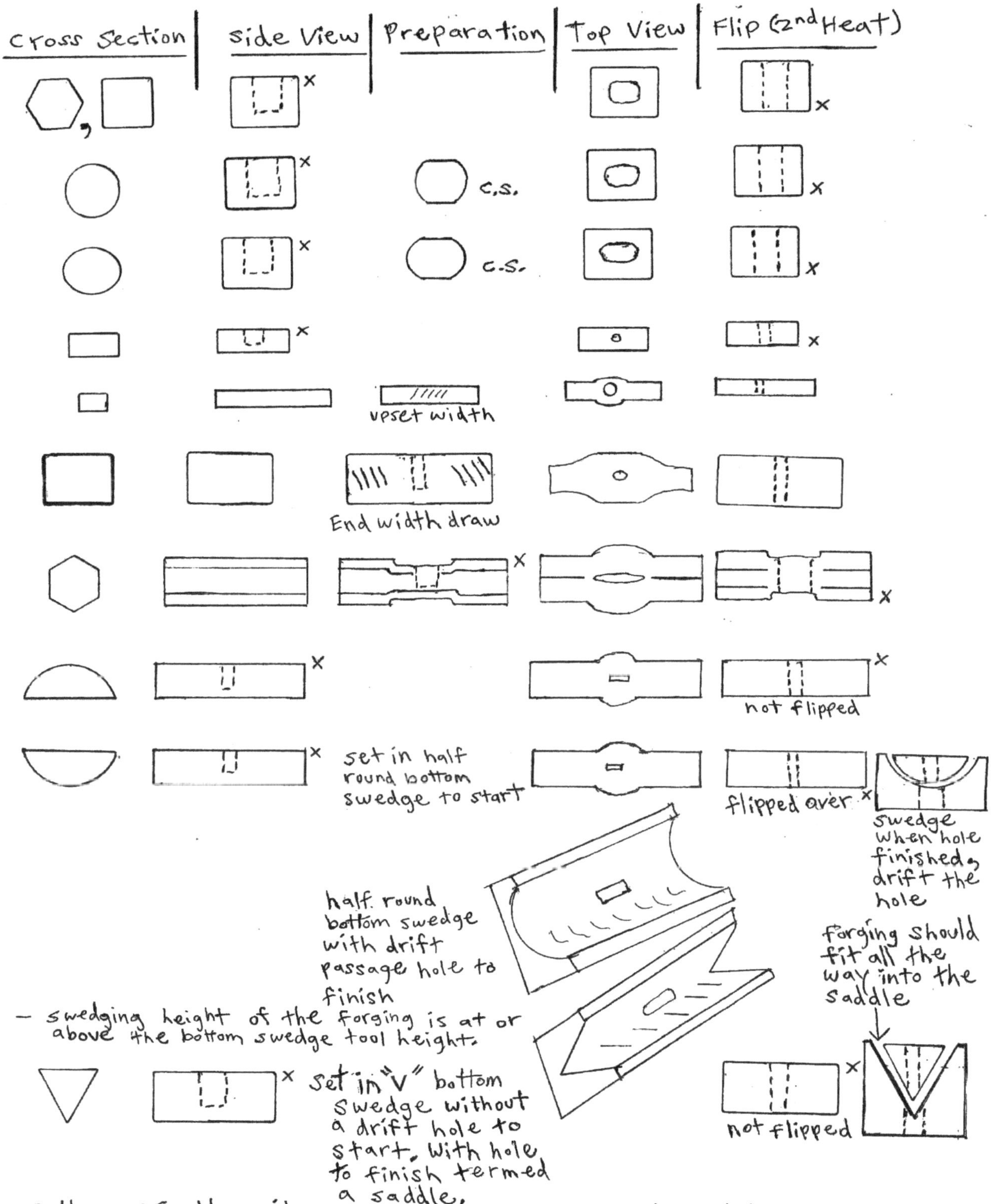

- Hammer the sides each forging as a punch tool becomes lodged; on the eye bulge.
- Many combinations of steel cross sections: punch profiles are used.

Drawing a Tool Brace (eye sides)
Under power hammer or with striker

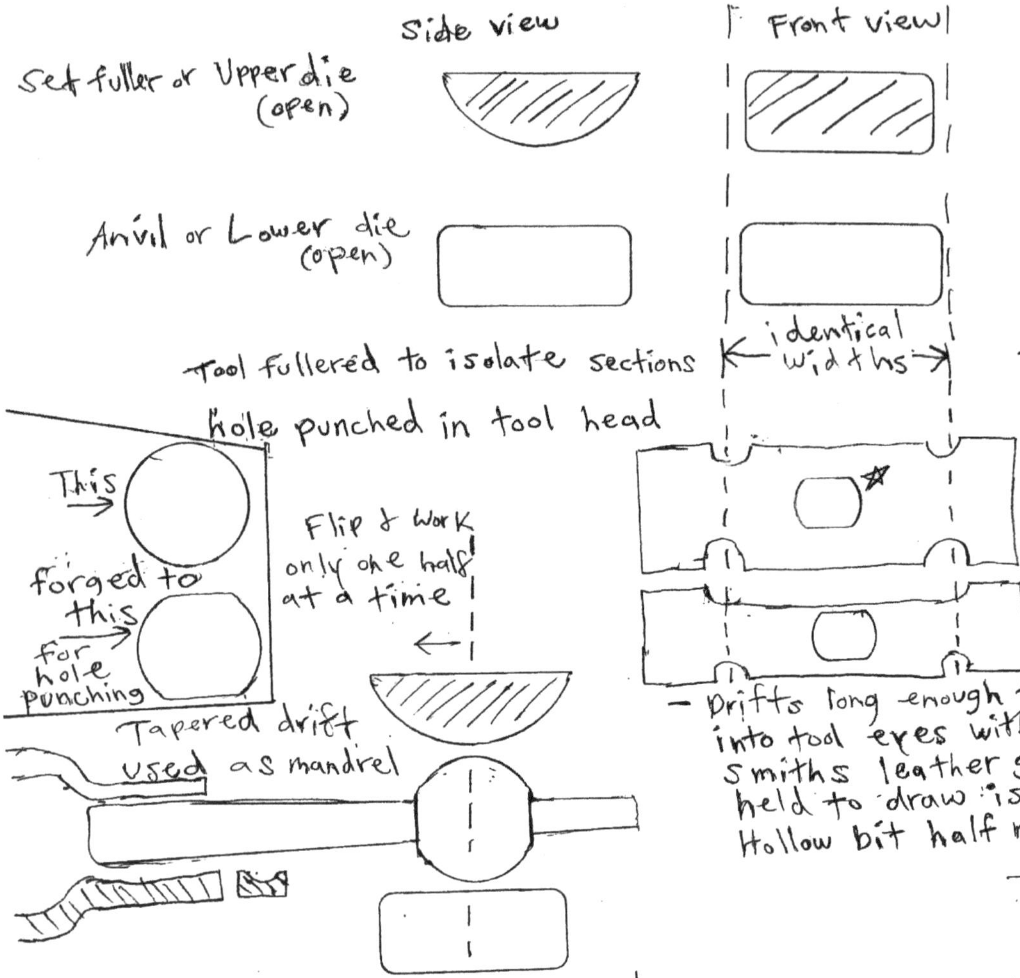

- Flat bottom die helps keep the forging still & safer
- Progress is swift and dependable

☆ eye shape important... no square edges; flat sides upper & lower are best.

- Drifts long enough to hand hammer into tool eyes without scalding smiths leather glove. Tong held to draw is best. Hollow bit half round or "V" jaw.

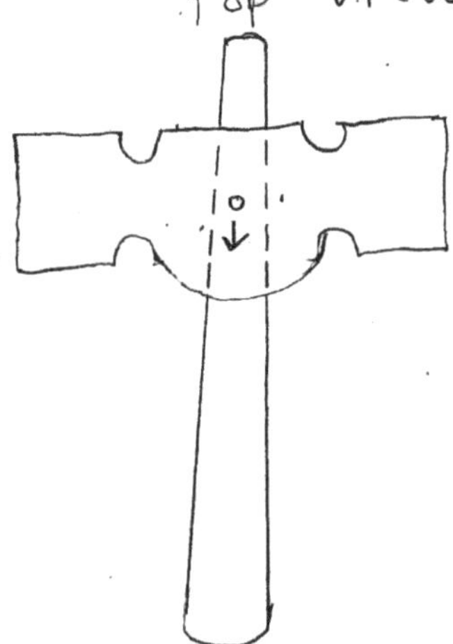

o Impact here & up drift

- Proportionate size of tool head to mandrel diameter shown...
- Paramount attention to time of mandrel in hole during drawing Extract before color of heat develops.
- Drifts allowed to air cool if made of air hardening steel before reuse.
- Drifts of carbon steel water quenched at a black heat. Bees wax greased.
- Hand hammer the next drift in opposite hole side. Draw to mandrel thickest end only...
- Have many drifts, progressivly thicker,

- Flip tool over to draw underside brace equally
- Take another heat and cooled drift to draw opposite exit hole

How our Anvils are Forged under Trip hammer Power

Steel Dynamics

material: salvaged wrist pins, track pins

mid sized anvil 100- 150 lb

Anvil Top

3-6" round

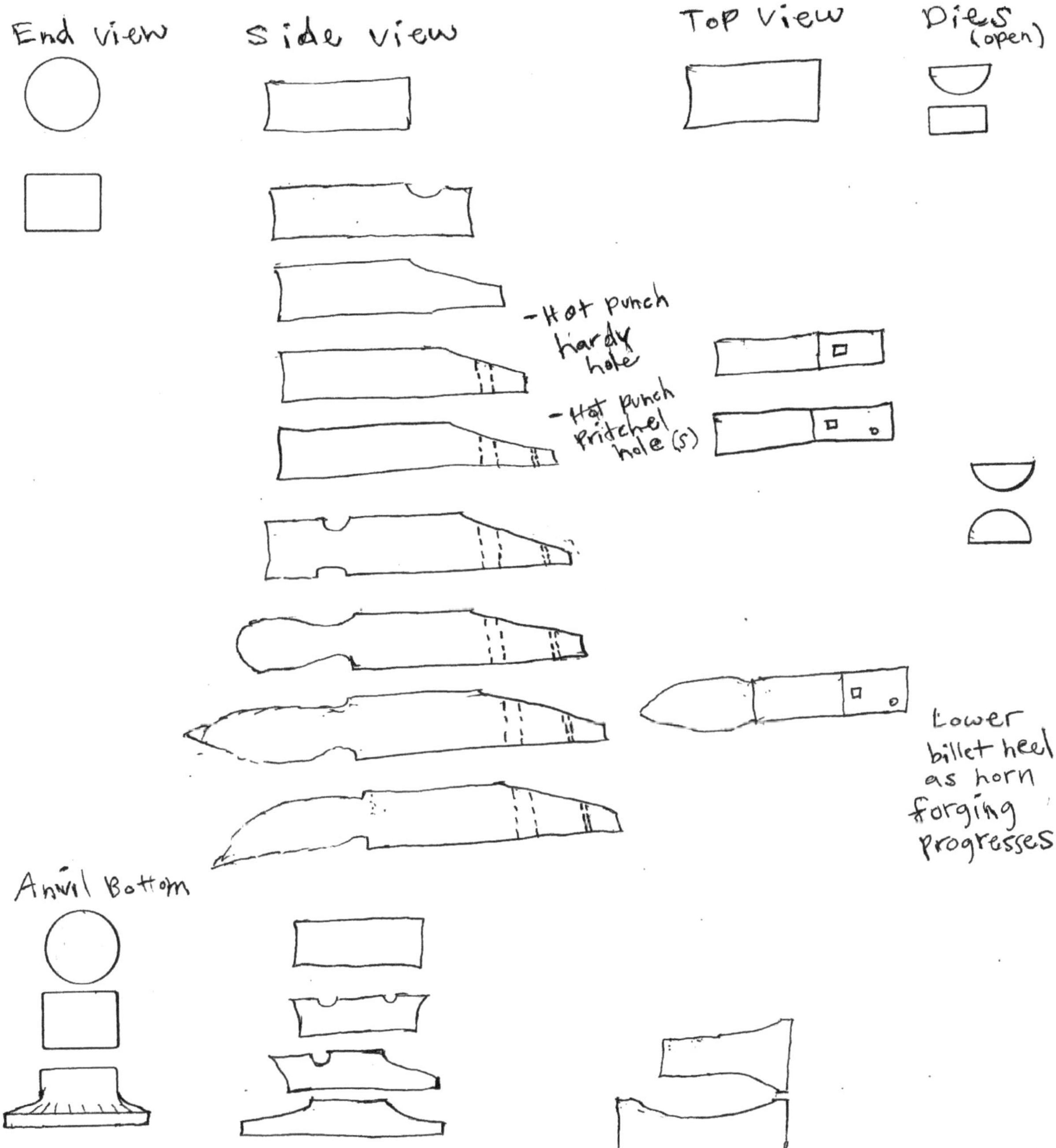

Anvil waist squared up block (separate)

To Join:

method1 - Fire Weld → Heat treat

method2 - Heat treat separates → arc weld to join

The Lock
A Simple Mechanism Design
Simple Purpose

Materials: mild steel – 3/16" plate (A36), latch
- 5/16" round stock
- 3/16"–1/4" round stock if hot punched for closure pin holes
- 1/8" round stock or 1/8" x 1 1/8" dome head rivets for closure pins and hinge pin
- 1/2 x 3/4" flat stock (surround spacer)
- 1/8" plate, mechanism cover, key hole cover
- 1/8"–3/16"–1/4" brass or copper plate for fancy exterior if desired

Spring steel – 1/8" x 1/2" (5160)

The better the steel, the more durable the lock

Pieces and Parts: Hot punched and forged holes work fine. Drilled holes can be slightly more accurate. A combination of each is great... We are building functional hardware, not a time piece. Artistic flair will definetly show through here.

- First Forging: surround spacer – Determines the overall body size of the lock

eg. 1/2 x 3/4 x 7" turned edgewise

Empherical (not to scale) relative in proportion

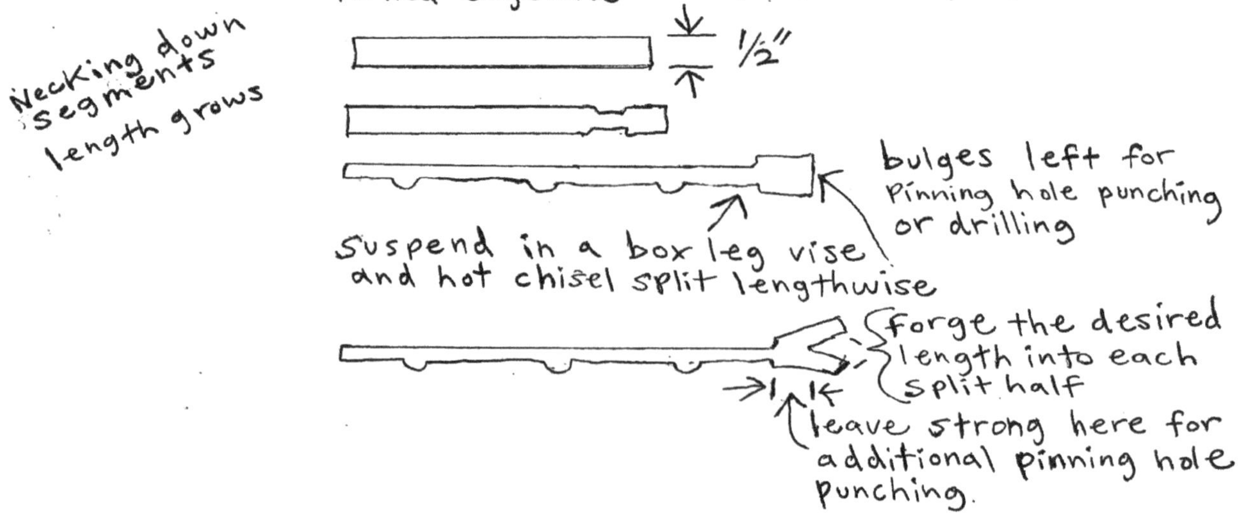

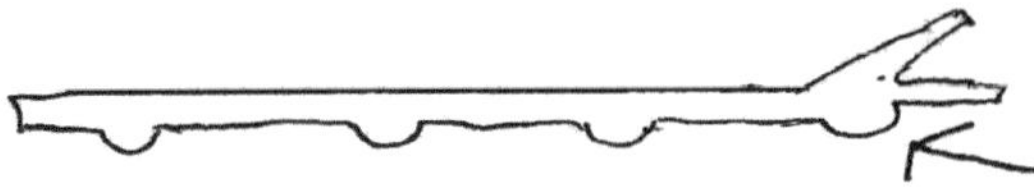

Prepare hole boss & offset shoulder

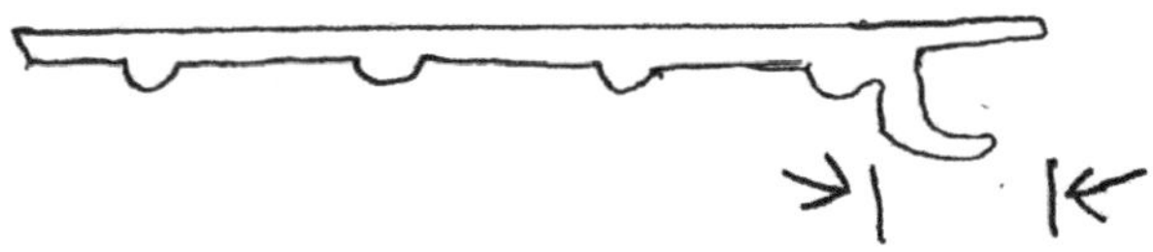

Finish drawing length to splits. Forge to this shape for latch hook entry.

1/2" 5/16"
thickness

- Bend and forge the surround spacer to approximate the final lock body shape desired. Remember that thinner segments will try to bend sooner than thicker portions. After the shape is approximated, lay it flat on the anvil and hammer all to level.

Crispen corner and bevel for hook entry.

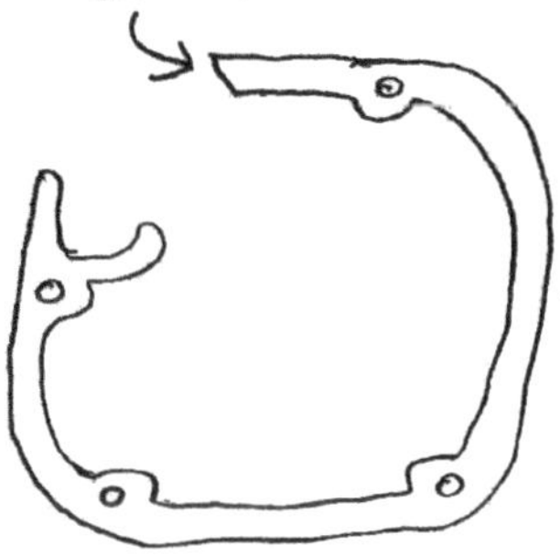

- We are now ready to punch or drill the through pin holes. To hot punch, do this just like a rivet hole. Drive a tapered round punch 2/3 through the thicker boss areas alocated from previous forging. Hammer the hole sides to extract the punch. Flip the forging over and drive the punch back through and clean out a nice hole. A small button of steel should have emerged from each hole made.

  Bees wax grease the punch working end before each punching process commences.

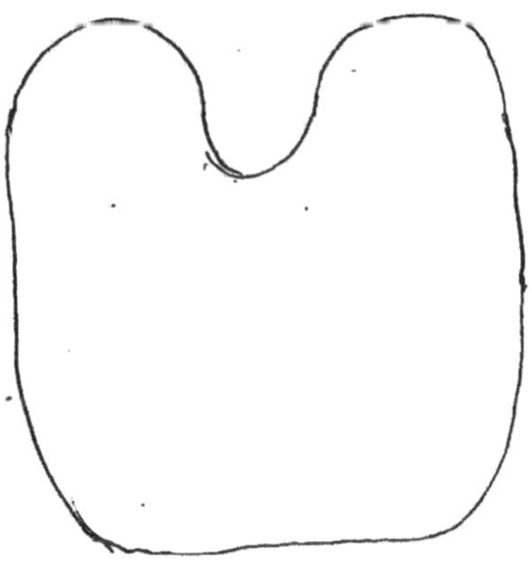

- Front and back cover plates can now be cut out using the completed surround spacer as a partial template. Once the external perimeters are traced on a plate they they can be cut out.
- To cut out: Torch (oxy/acet.)
  Forge heat and hot chisel
  Band saw
  Punch press (hot or cold)
- Ragged edges exposed are ground clean and given a uniform radius.

– Back plate with surround spacer superimposed shows where additional holes need to be punched or drilled.

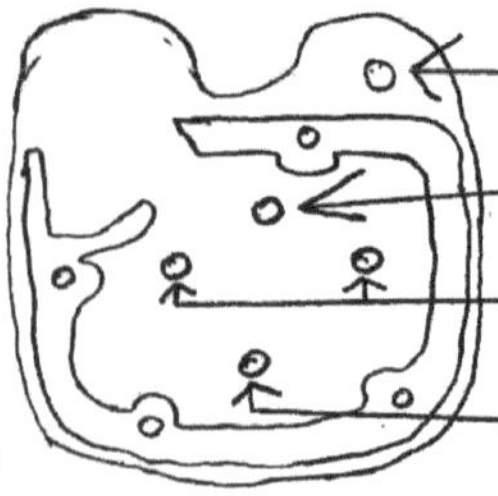

– Front Plate

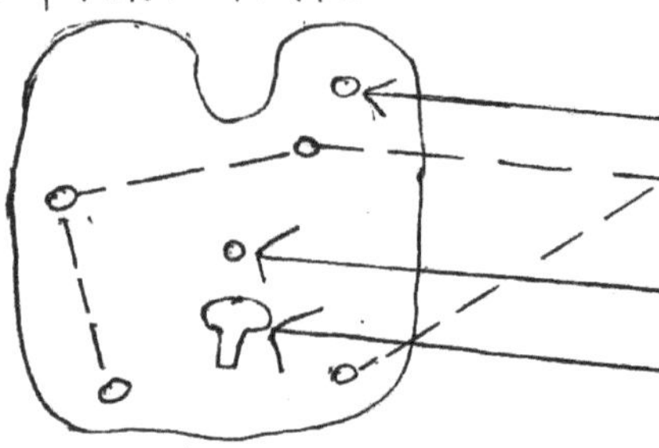

– Key hole – cover plate can be any shape

– The mechanism

– cover

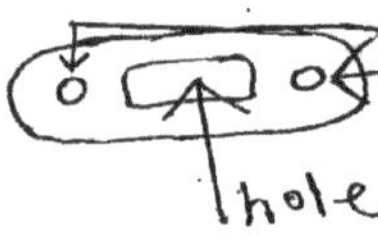

Pinning holes for attachment to to back plate. This plate is stacked to be closest to the inner front plate portion of our moving mechanism.

– spring

only one pinning hole
The spring arc position is adjusted for correct leverage tension after all parts are made. Allowed to air cool after forging. No heat-treatment. Careful not to overheat one during forging and hole punching.
☆ Latch keeper nub protruding into hole. This is the nub which keeps the lock locked. Regardless of its position.

– Latch
side view

End View

The latch is the only part with oblong mount holes. Pinned snug but free enough to slide left and right when asked by the key.

The Top Hook

The hardware which is exposed and captures whatever item requires security.
Our tools, whether purchased or handmade and used are precious. Quite often not replacable. A good lock and sturdy top hook can save much heartache.

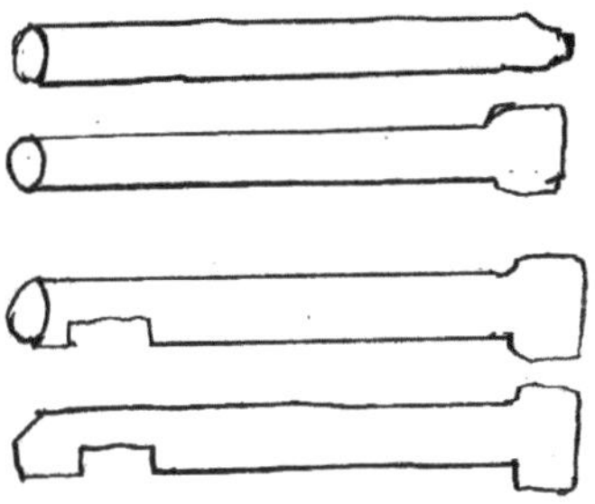

3/8" X 4 1/2" round-End corners forged down.

Heat, differential quench, bump up the hinge pin end

Forge and hot rasp the catch end and bevel. This makes for easier passage and entry to the lock body.

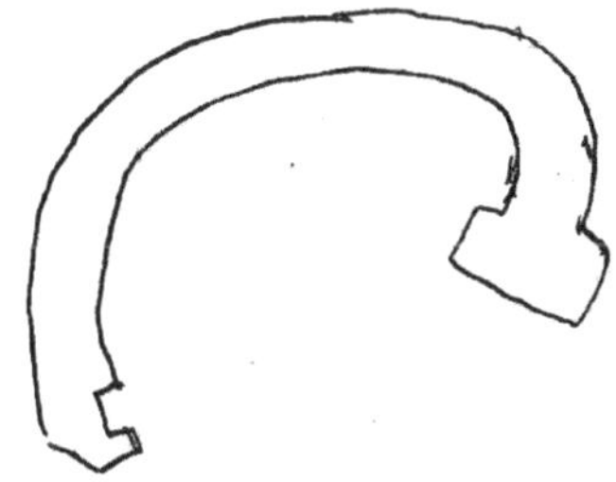

Forge the top hook to a shape similar to this; depending on the body shape of your lock.

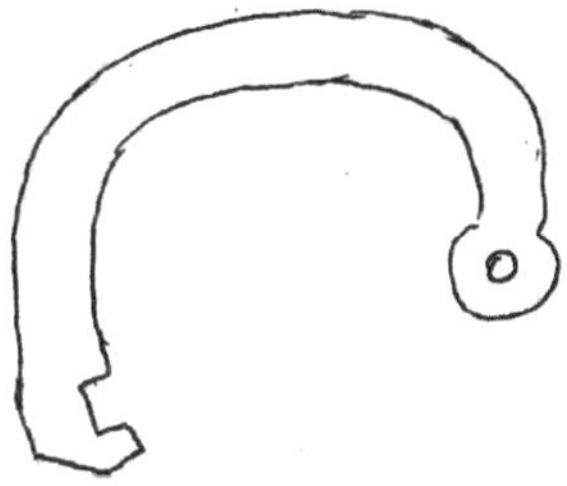

Now the hook will lay flat and hold still for the hinge hole punching or drilling if preferred. Flatten the hinge end slightly. This will yield a greater surface area to send the punch through. Also, the hole sides will have better mass. Round up and flatten the hole bosses as needed.

The Key

Any shape which will push the latch movement recession. Provide lateral direction to the latch after entering the front plate.

Two key types here;
with pipe hole
with an end nub

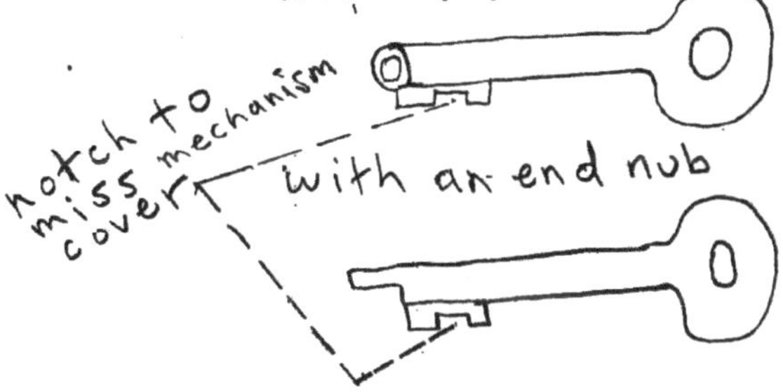

Hollow pipe/tube slides onto back plate key pin. This key pin is installed to protrude only 1/2" into the lock body.

The back plate key pin hole is left open and serves as an entry and pivot point for the end nub.

Gather up all the parts. Give them a final inspection, dressing, dry fitting and start assembly.

Put the Lock Together

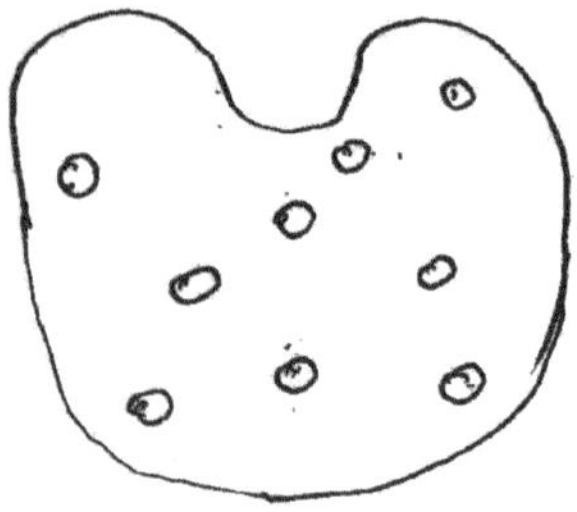

- Back Plate

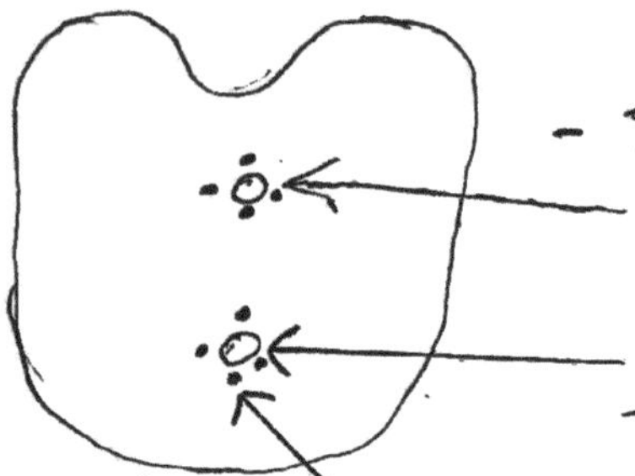

- Install these two pins first
  spring catch pin

Key pivot post
If a nub pivot Key will be used, this hole remains open (postless)
These pins do not protrude through the front plate...

here

After peining them over on the back side of the back plate, center punch inside the back plate right next to/adjacent to each post. Now the post won't move in or out of their back plate holes.

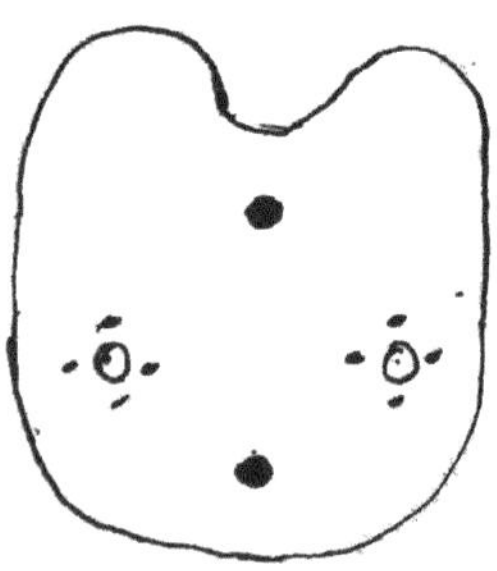

Next install the pins which retain the mechanisms moving parts. Again these posts do not come through the front plate and are center punched on inner back plate to lock in place.
Four posts are now in place

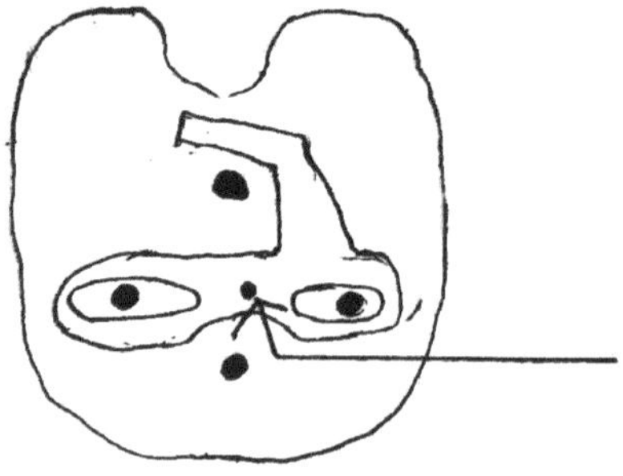

Latch placed over posts first.
Oblong holes allow for left and right movement (locked → unlocked).

The latch has a post of its own protruding ~3/8" toward the reader.

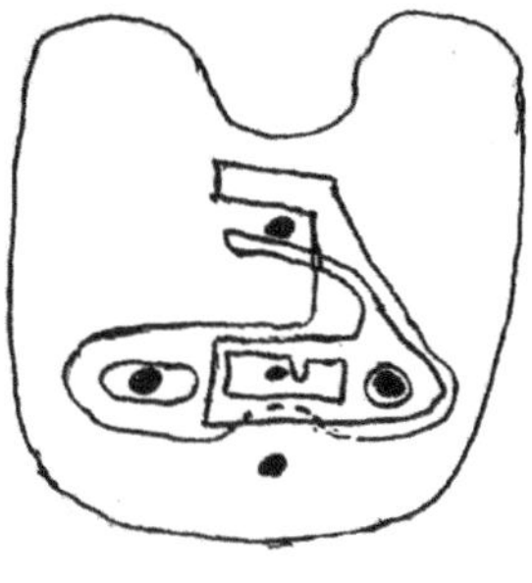

spring plate placed over the mechanism post to the right, on top of the latch. The center post attached to the latch protrudes through the spring larger square hole with nub.

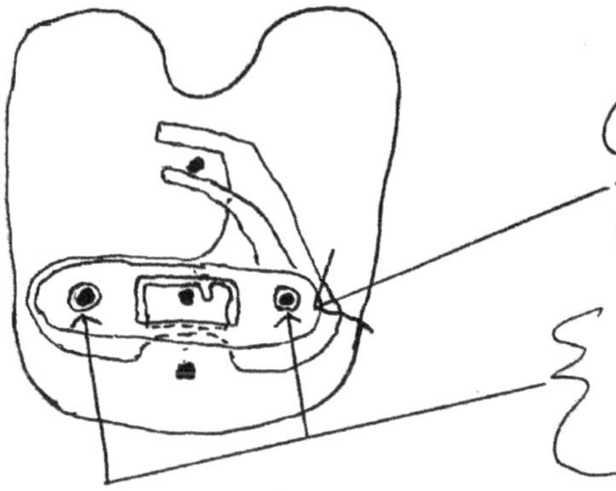

The mechanism cover is now placed on top of the spring and over the latch posts

Pein the tops of each mechanism (2) post over to lock the cover plate in place. Done cold with a center punch or set rivet heading tool. Tool selection for this depends on on your post length and diameter. Pick tools which enable peining the post ends without bending them. Hammer control is important here!

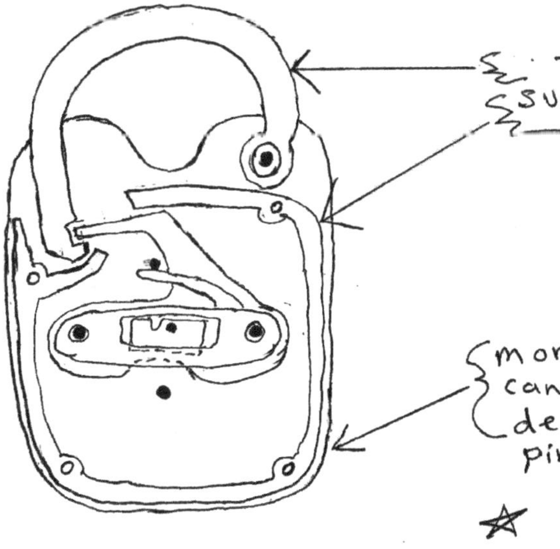

Time to put the top hook and surround spacer in place for pinning.

more surround spacer pinning holes can be put in a larger lock if desired. Space them and increase pin sizes accordingly.

☆ Check the mechanism action with the key before covering and attaching the front plate. The key is ground / filed until it moves the mechanism smooth and positively.

Locked Position with mechanism exposed. (no front plate)

Draw or attach ***your*** favorite sketch here.

Drawing – page for practice here.

SIDE AND WOLF JAW TONG DRAWINGS

SIDE AND WOLF JAW TONG
ORIGINAL DRAWINGS

Drawing 1

Rivet for long material holding tong

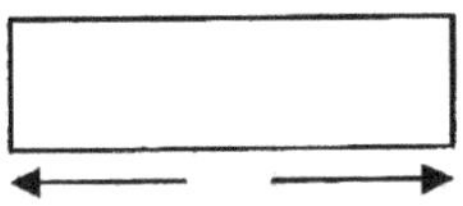

1/4 x 7/8

5-6"

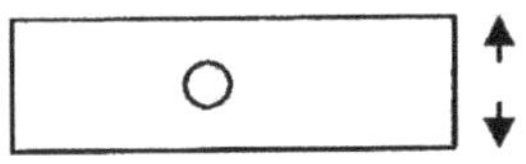

7/8"

Rivet hole punched

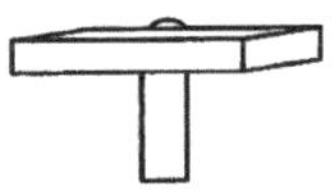

Rivet inserted 3/8" diameter of shank.

Weld on box leg vise or anvil edge or drop in pritchel hole

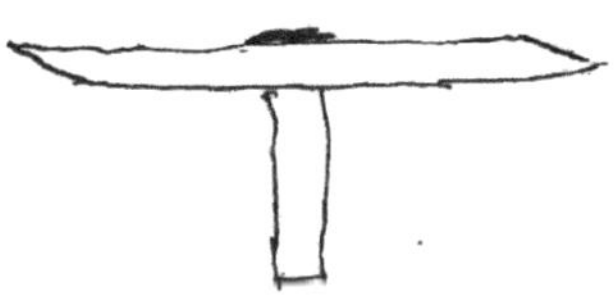

Scarf ends of strap

Bend around and forge weld on horn.

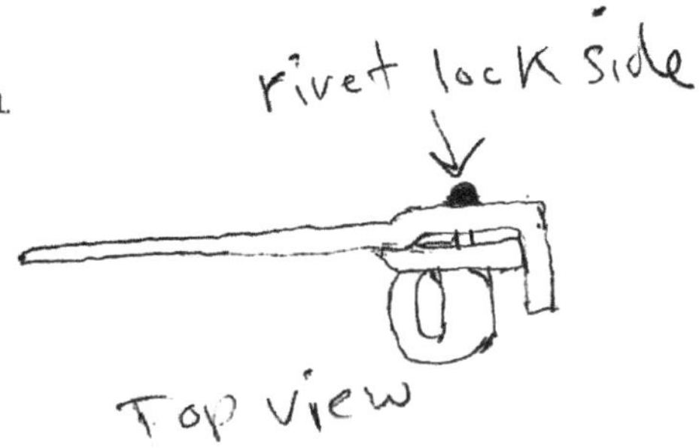

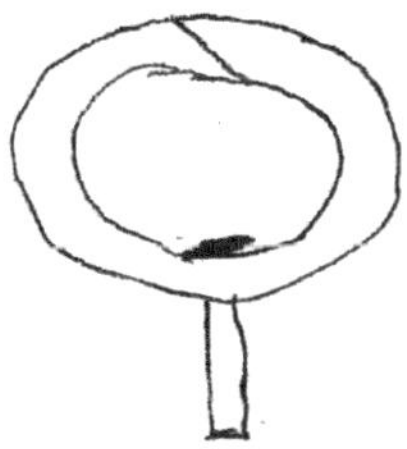

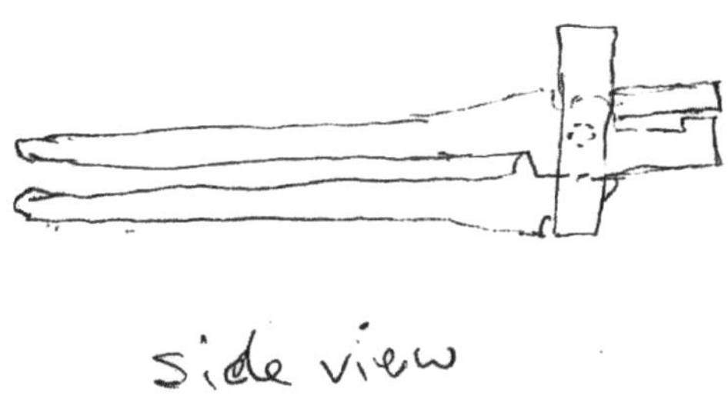

How to align accurate gripping tong such as wolf jaw, pick up tong – single and multiple size hold

Drawing 2

Wolf jaw

Rivet hole punched before tong jaw is grooved (jaws are still blanks without definition).

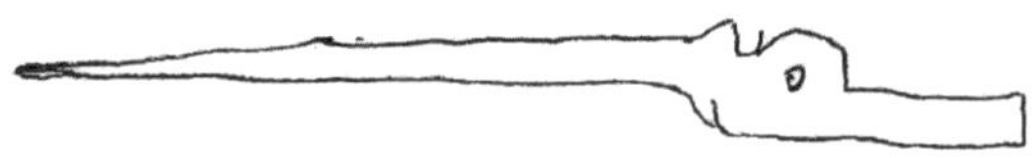

Top view

Rivet holes punched in both halves of blanks. Some may choose to drill their rivet holes. Hot punching rivet holes is preferred as the boss material is preserved and displaced to maintain strength (width) of bearing surface. They can later be drilled and/or reamed with a slightly oversized bit to true up the entire hole. This can be done with a drill press or large hand held drill. The chief requirements being to secure the tong blank in a vise or on the drill press tool rest. This is to prevent the tong blank from grabbing the drill bit and becoming an airplane propeller. Safety is placed first.

Drawing 3

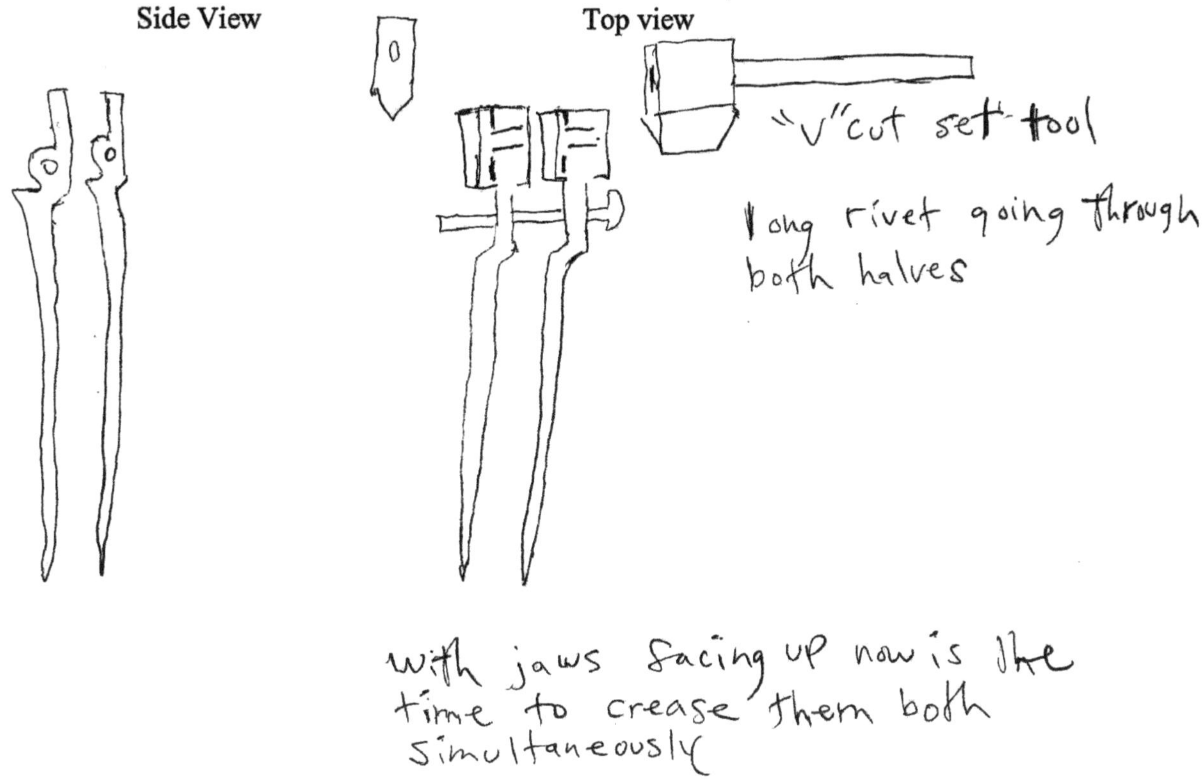

An extra long rivet inserted through both halves of tong which are placed side by side; not in the position in which they will operate - just side by side. This rivet is the correct diameter of shank for the hole punched in the tong.

Now we have a reference point, or as in surveying, called a bench mark from which to continue our forging. This rivet hole will help us to ensure that each blank will be marked exactly as its working counterpart. Place both halves in the fire to heat the jaws. At a good orange heat, extract them from the fire together. Only the jaws are hot at this point.

Held side by side on the anvil insert the extra long rivet through both halves. Now is a good time to scrub the billets with our wire brush to remove scale and slag.

The blanks are now marked simultaneously side by side with a hot chisel. All we need is a shallow hot chisel mark as few or as many as desired by the smith's needs for this pair of tong.

- The rivet is then shaken out onto the anvil to cool.
- Our blanks are put back into the fire to heat only the jaws.
- If an open mouth is desired on the jaw, now is the time to mark it with your hot chisel.

Drawing 4

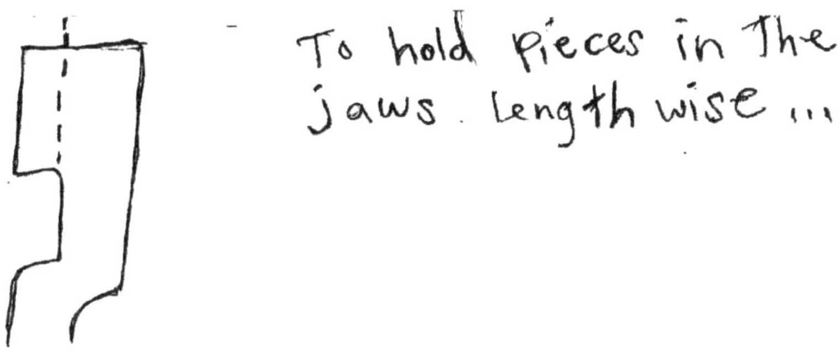

This is done one blank at a time. The mark made to meet with the bearing surface side of the tong (throat to lip) inside portion of jaw (top view). Each blank is marked the same way.

Each blank is heated and the gripping jaws are driven with top tools to their desired shape and depth. The location has been well taken care of. We just need to finish the forging.

For each blank it is important to drive each jaw groove to the same depth as its opposing jaw. This will help ensure each groove is equidistant from another. We are displacing steel now; giving equal forging depth to each jaw will keep them equal in form and function.

Drawing 5

End view of lips on jaws

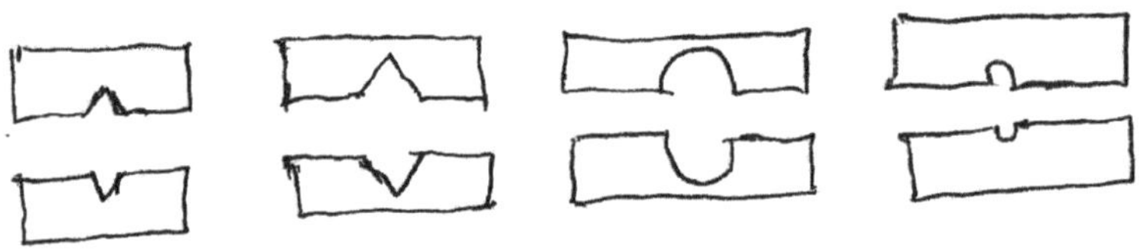

Side view of lips on jaws. Many variations are possible and used in our smithing. This includes mass of the jaw, length of reins, number of teeth, cuts per jaw.

Drawing 6

V cut wolf jaw

Fuller cut wolf jaw

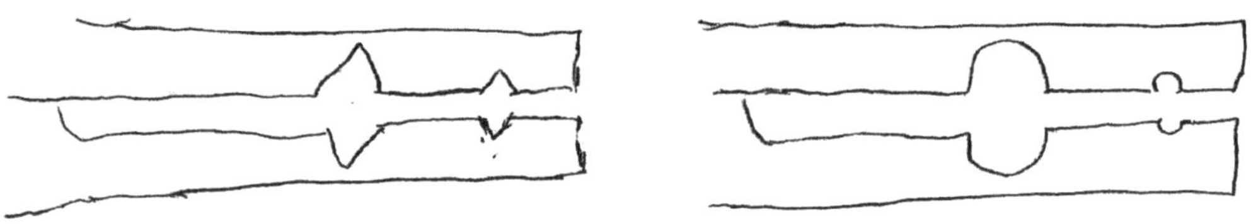

A V cut wolf jaw tong will hold square and round stock very nicely.

To complete a nice V in all parts of the jaw, a top tool will have to be made.
Drawing 7

End view wood handle.

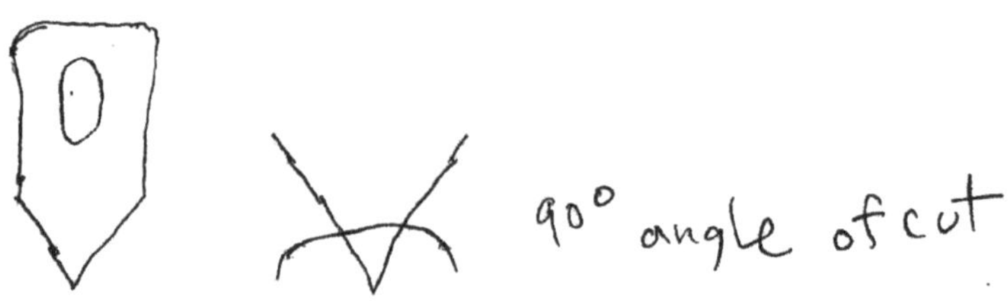

A good tool to have for it can be used later for forming hollow bit tong.

Drawing 8

Top & bottom used together

End view

this

turns into

TENSIONING

TENSIONING

This term refers to the use of the steel project itself to hold the successive part to be attached in place. This is used quite often where gas, arc or even forge welding is required. The next part to be attached, regardless of the project size, is cut slightly oversized. The base metal of the project is checked for the amount of spring or recoil tension it can provide when tried. The successive part to be added is measured and cut a fraction of an inch oversized. When base metal is spread and this part, just cut, is placed in position; it is not only accepted but held in place for welding. Because we are a single man shop much of the time, this tensioning method is employed quite often. It has made a third hand unnecessary even with large projects.

<u>Weights</u>

Weights are often utilized to allow our steel to maintain its set up position.. One to five pound chunks of steel material. The shapes of each are as varied as the positions in which they are laid to rest. Much effort is made to place them carefully so as not to obscure our view of any material held beneath. This is often preferred as opposed to clamping.

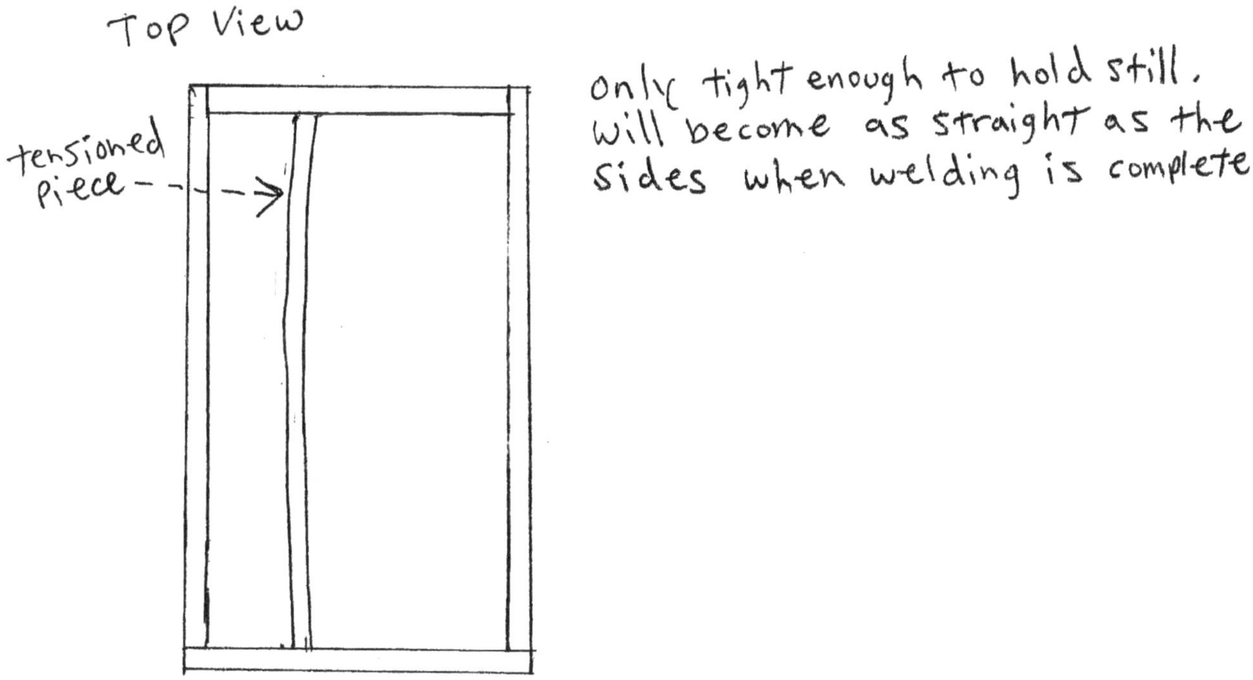

SKILL

SKILL

Over the years, there have been many conflicting definitions with regard to skill in forging ability. Many attempt to shed light on a smith's eye/hand coordination. Some are more determined to attribute much success with experience that dropped from thin air. Others seem most preoccupied with the amount of material left for grinding and finish work. Each presents an excellent argument. Accurate and intentional hammer blows do equate to less time spent making corrections, hence greater hourly productivity. If the smith is to prosper, production costs must be kept reasonable. Minimizing losses in finish work is good. Therefore, the forged product needs to be as close as possible to a finished unit when shaping ceases. This can be termed a skill. The residual material may require some small amount of grinding/polishing to become uniform or a marketable product.

Most often in open die or free hand forging, the skill demands for repetitious or mass production for identical units dictates a high skill level for the smith. When similar items are needed, the demand for excellence can run high. Attention levels can become sullen or even border on the mundane in multiple run items. Active and passive attentiveness and also high levels of skill are to be continually employed and mastered.

If we can run our mortal machine at a steady cadence who knows what skill levels will evolve. Active and cognizant participation demands that skill levels will improve. Where does this skill come from? Answer: PRACTICE.

A mental basis for power which is dependable through our hammer swing should be sought after. A keen recollection of our hands and their manipulation abilities. How quick they can turn, lean, pull or push a pair of tongs? How steady can a piece be held during its forging? Some smiths may seem untouchable for their skill and fortitude with items that are turned and drawn incessantly. Yet for material that must be held still and struck in one or two specific segments, this same person may falter and seem disenchanted. Not that skill cannot evolve to bring up a weakness, but the inclination is not within their personal preference.

We are all good at some aspect of our work. Let's find out what that aspect is and add to it many attributes. Remembering that a chain is only as strong as its weakest link, why not add accuracy, precision, experience hence over time and skill? With this cognizance in our daily practice, dependability, fitness, reliability and efficiency will all evolve as well.

To become well rounded, both strengths and weaknesses should be practiced. Each practiced will enhance the other. Attempting different types of projects helps our hands become more connected to our eyes. The eyes will learn to see what is missing. Learning to recognize what is already there and could/should do best left alone; maybe only slightly modified. Essentially a practice in repetitious vs. static or single blow forging projects will provide a tremendous gain in experience. Now our mind doesn't have to wear out the body to find a satisfactory finished product. Learning which forgings require many hammer blows and which don't. Try to recognize which venue is most logical and achievable within or at the far reaches of our abilities.

The smith's experience through repetition is now elevating his/her level of skill. For items that are one of a kind, the judgment concerning where/when it is called finished is left wide open. Our goal may be just to find what would be called satisfactory. That is fine for most of us; when the eyes receive a picture and can tell our hands that enough of our heart and soul has been imparted to the work. The skill here is knowing when to stop. The project became what was intended so we can call it good.

We have heard it said many times that "experience cannot be taught." When some conscious effort is made to impart specific virtue to a forging, the slightest bits of skill improvement will unfold. It may seem minuscule at first, but the foundation for reliable skills are emerging. Over time and practice, so does the experience.

While heading for a fine run at a forging, take a moment to see where you are. Take a gander or measurement of what is done so far, and is it shaping up as was desired. Developing the thought process to stand up once in awhile and catch a good look at where things are in a forging is a skill many forget to acknowledge. Impatience will catch up with us all at some point. Patience is legal and a virtue for most all. Developing our eyes makes for progressive forging and less correction of mistakes later on in the work. This is a skill unto itself.

STORY TIME

STORY TIME

One of the best blacksmiths we have ever come across had quite a surprise to share. After watching him for years, trading secrets, conversing and working together he approached one day during a tool making segment and said, "Can you help me draw the colors on this punch?" We were tempering a horseshoe nail hole punch. I asked him "What's wrong?" Again he said, "I can't see the colors; I'm colorblind." Well, after almost picking myself up from the floor, the thoughts were racing a mile a minute. How could this world class great fellow have gone so far with such a handicap? He had taught himself to use every means possible to make sure steel didn't become overheated in a coal/coke fire. While watching for sparks and trusting his eyes to use only that as a gauge for appropriate heat levels. The same for a gas (propane) fire is very possible. Talk about teaching our eyes to become a true window to the mind. As the steel moves under the hammer, every sense of its heat is recorded by the amount of movement exhibited. The skill is practiced and elevates to where hands, tong, and hammer feel where one step is completed and heat is diminished. Please take a moment to ponder this. . .

The smith can then develop good judgment for each heat. Is there enough heat in this billet to bridge the gap from where it is to where it needs to be? Can we get closer to the finished product desired? When our feel for mass of billet is working. The force from the hammer we can impart to it; whether under power hammer or in hand. Combined with the degree of heat we can provide to it, a recipe is formed. This recipe is different for each smith, though a maximum and minimum for each ingredient does exist. With practical searching for this recipe we can learn to operate with this realm. Keeping it closely in check and not to proceed without it. Later, forging doesn't feel right without it to where we wouldn't even consider forging without its involvement. Balancing the heat versus forging power recipe is a skill. Our senses tell us from experience when there is enough of each present. We can then feel the steel move in a fashion that comforts our persona. Our eyes can keep up with the progress. No excess force is required. The full elastic properties of the steel are then utilized. Now the smith is beginning to find a comfortable zone in which to operate.

A beginning/experienced heavier hitting smith may prefer cooler heats. At first, the eyes can see the progress occur more easily, though there will be less of it. There may be more comfort in this as inefficient as it is. The eyes can more easily track the progress in slow occurrence and adjust the mental picture for successive heats. Slow but sure is the realm. The more svelte or lighter hitting smith may require more heat in the billets to achieve a softer, hence more forgeable material for them. This smith needs full bloom or hotter billets to come even remotely close to this comfort zone and have efficient results in the hand hammer blows delivered. The eyes must work harder to keep up here. Once this type of feel pulls its recipe together, the eyes may be the only factor holding things back. The old adage has it that the hands are quicker. For forgings requiring finesse to draw, and/or fold, the smith may have to just slow up, breathe, and let the recipe cook. Most steel can only be heated so hot before it is ruined. Conversely, when too cold it will not move efficiently toward a finished product. The amount of force delivered is metered and rated through feel as proportionate to the amount of displacement possible for the forging. This is where the feel for balance in your recipe is so very valuable.

To gauge this, another sense is developed. This portion of our attention ensures that steps requiring much bending, forging or reorganizing are attended to when heat is in abundance. Simple alignment or corrections are made as heat is dissipating. From an uninvolved observer's standpoint, we may think that heat loss occurs consistently from the time it is extracted from the fire until cool enough to touch. Yet under hammer, a given section can be made to carry a forging heat for double to triple the dissipation time. This is the realm we are teaching our mind's eye to recognize in the steel and utilize. A good test for this is having two similar billets heated and extracted from the fire simultaneously. One is hammered, the other merely set on a fire brick to rest. Take note of the time color from heat runs out of each. Learning to identify the heat developed from hammer impact, friction and compression recognition is a skill unto itself. For most any size billet being forged, this heat developed from impact can be measured both quantitatively and qualitatively with our eyes and ears if we are looking for that answer. This can and will become a very dependable asset in our arsenal.

As with all competitive events, there is a common denominator, the quest for excellence.

While repairing a branding iron for a fellow, an interesting discussion in perception arose. The brand looked like this:

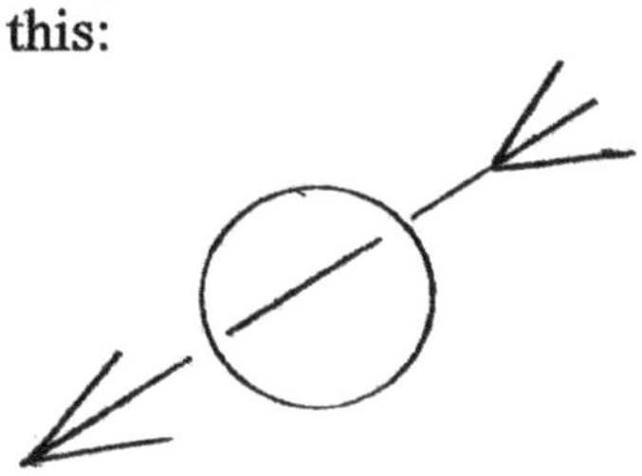

Drawing

He called it a circle with an arrow through. Some may see an arrow through a circle. I wasn't in disagreement with him, but the brief description left much room for cognitive fuel. The point being, as a handled tool to be built is visualized, what do we see? A hole with a tool around it or a tool with a through hole?

As design dictates, many see the tool first and later putting a hole in where convenient. For set tools this is fine (tools that are rested then struck). For swinging tools, such as sledges, rounding ball/cross peins, my preference is to look for the hole as a first priority or decide where can the hole be put. Hole placement is paramount for balance, hence usefulness.

What does the tool need to do? As a tool design is being created, my tendency is to work backward from the handle. To insert a wooden handle, it needs to have a hole. Take the cross pein for example--a swinging tool. While looking at the side view of one on paper or in person, we cannot see the striking surfaces nor the through hole. Without seeing through the hole, we cannot tell if the hammer is balanced end to end (one striking surface to the other) when the handle is inserted it becomes apparent, the hole and handle must bisect the midsection of it's mass to create balance. As the hole is punched closer to one end or the other of your tool, a different sort of balance is created.

Top view: Where would you put the eye to balance these hammers?

side view

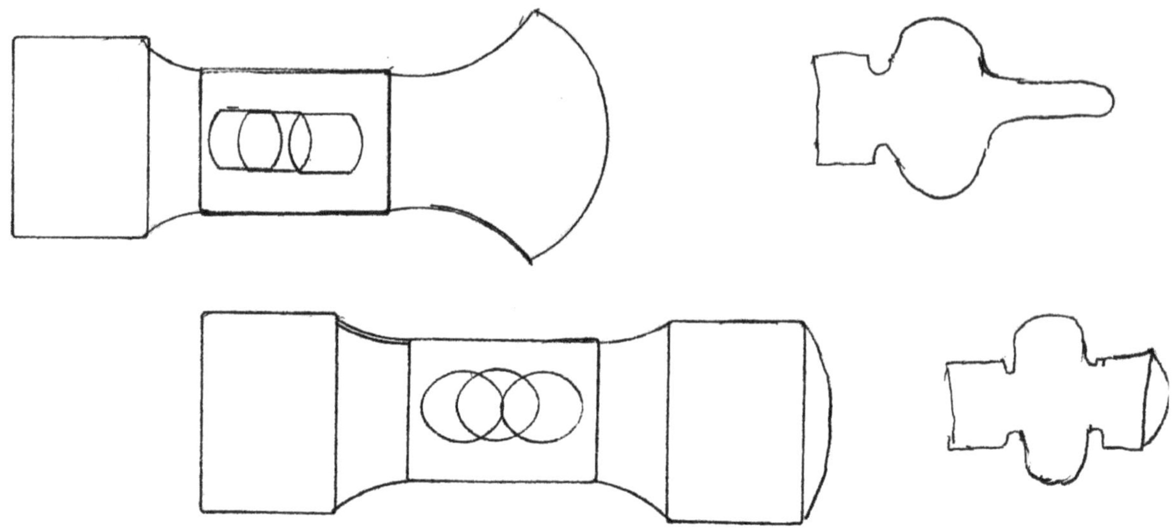

What creates balance for a swinging tool? For our purposes, equal weight on either side of the hole creates balance. This applies regardless of overall hammer weight, type and distance from the hole the striking surfaces reside. A hammer used to simply rise and fall for leveling material can be heavily offset. Flat surfaced hammers for leveling or straightening items such as swords and knives can feel very useful, but these same hammers used to forge the sword billet, forget it. More strain on the arms and back than any smith should endeavor to endure.

Set tools feel most comfortable when there is more weight on the working side, between the handle through eye and the present work. They hold steadier where placed. Not top heavy or too much weight between the handle and striking surface. These are difficult to use from any head weight, including ounces to tens of pounds. Set tools whether punches, hot/cold chisels, flatters, creasers, veiners, etc. are most comfortable, accurate and manipulable when their center of gravity is closest to the work. There is much less chance of mis-hitting work because of weight distribution principles.

SWEDGING

SWEDGING

To create a smooth handle on tools such as nippers and nail cutters much consideration should be given to the rein profile, half round to the outside portion of the rein which touches the users' hand when in service. How is the appropriate rein profile achieved? Since most all rein profiles are drawn out in a forging square (open die forging) how do we get to 1/2 round. Is it then put in a 1/2 round bottom die or swedged with a top tool? My experience has shown that when square or rectangular profiles are achieved in the work or project billet, a concave half round top swedge is the best bet. One that fits just over the top corners of the billet and allows forging down to leave a nice 1/2 round crown before bottoming out. Bottoming out meaning the top swedge can be driven into the billet to a good depth before its lower-most edges hit the bottom die or anvil beneath.

If the reins are forged or drawn out square to octagonal then round, as are many tools such as farrier pull offs or many tongs, another approach has been found useful. Often a round tong or rein profile is most desirable. Yet the surface area that enters our hands may deliver a concentrated shock from the tool's use. To help dissipate this shock or sting factor, additional forging may be essential.

Drawing 11
(tong rein profiles)

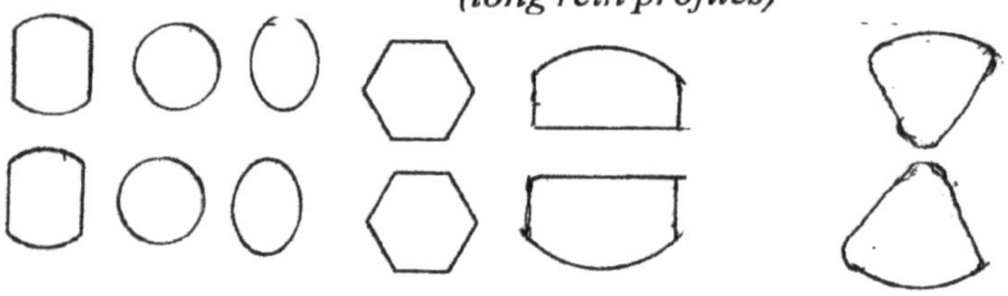

Drawing a rein square, then octagonal, next to round.

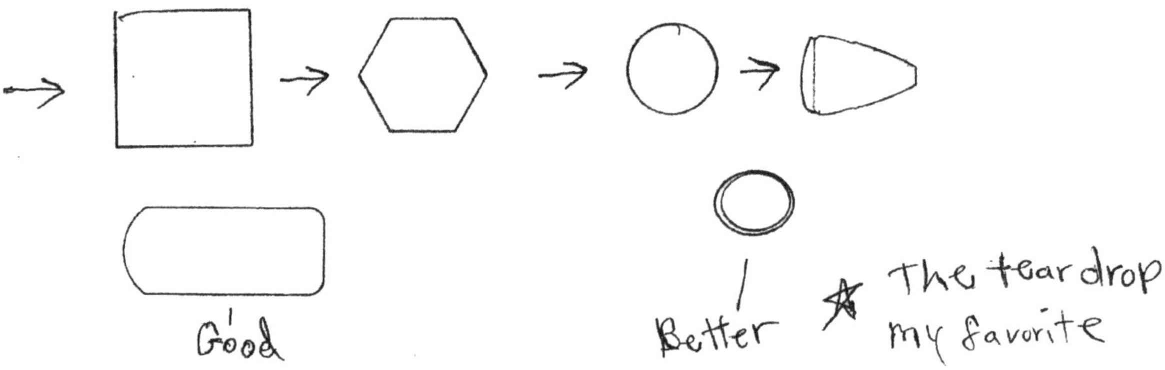

Now we have many more options open to us. On a round profile 1/4", 5/16", 3/8" or 1/2", I have found running this through a bottom swedge with a larger radius than the stock at hand creates a half round rein. This creates a nice wide contoured surface to present to our hand. The flat portion to remain on the medial side of the tool, contoured side out.

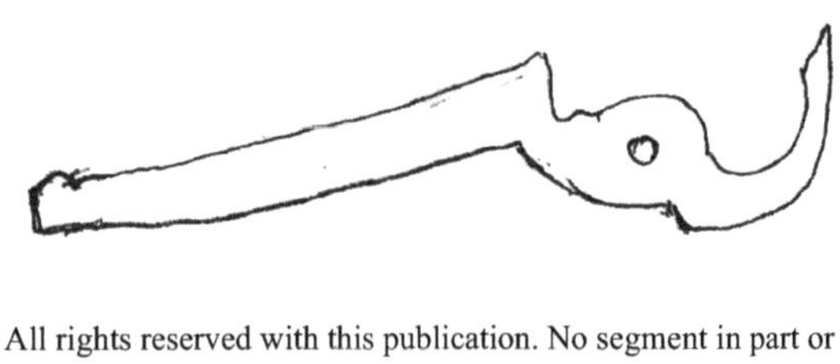

Drawing 12

(nipper reins)

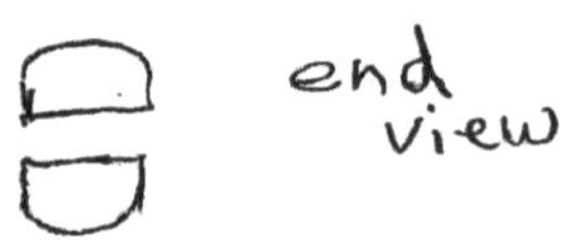

GOALS AND EXPERIMENTS

GOALS AND EXPERIMENTS

It is my goal not to just disclose some of the mystique behind our own methods of forging, but to unlock some real truths.

This is one of the doors we must go through to produce work that stands out from its predecessor, head and shoulders above previous attempts. If we use our eyes we can see far, but if we use our hearts and minds we can see even more amazing things.

Let your eyes tell you what there is. Let your mind's eye show you what there can be. Many considerations should be given to how material is handled for forging. Some parts should hold still and others must move. Where and how your material is held is often just as or more important than where it is hit.

This applies to both hand forgings as well as under powered machinery. When good positioning of work material is combined with accurate and controllable means of manipulation, the power proponent can become slightly more negligible. Yet if all these variables are helped to work in concert, remarkable and repeatable results can be orchestrated, helping our eyes and mind to remember and record. Which of these ingredients are working as was/is desirable is essential. Remember that for scientific study (as a controlled study dictates), to be reliable it must be repeatable. Under similar conditions, e.g., heat, time and position similar results can be achieved consistently. The smith's control of these variables can be monitored and regulated. It is our practice of these aspects which allows us to succeed.

Learning to observe, remember and compare where the project is and where it needs to be. Where it was and what the smith has developed it into. Which of the aforementioned proponents is working best for the blacksmith? Which could be working better? All aspects we, as smiths, must learn to monitor. At first it may seem a conscious effort as it should, yet with time we learn to check in on each element at a moment's notice. Help correct and fine tune the element in need. Over time this should become routine. Sounds simple enough, but these are the rules. Try getting them to work for you.

PUNCHES

Pictures 58 & 59
Close up of flat side of oval eye punches. Note that all of these have a prominent shoulder at the punch tops. This part is necessary for our extraction process, to remove them from initial holes during punching. Each working end has been bees waxed and is ready for future use.

Picture 60
These are round hole punches all used to make different types of rivet holes. Although varied in design, they each achieve an identical purpose. All can be used in hand or under trip hammer power. The wooden handles shown are of hickory and eucalyptus. Many fruit woods and osage orange work well for these set tools.

Picture 61

These are all different sizes of flat side oval eye punches. From left to right we show the smaller to larger diameter. Made of H13 and S7 air hardening steel on their working ends with mild steel handles. The height and heft of each has a different purpose. Finding success and perfect volume for these can be somewhat of a journey rather than a destination! Matching the right size (diameter) to punch a hole in a given size of steel will take practice to get good holes, but it will come. If done by hand hammers, a smaller hole is easier made first, then drifted out to larger ones. The handles should be light enough or have a narrowed section to relieve shock transfer to your hands. The height should not be so great that proficient and deliberate hammer blows cannot be delivered.

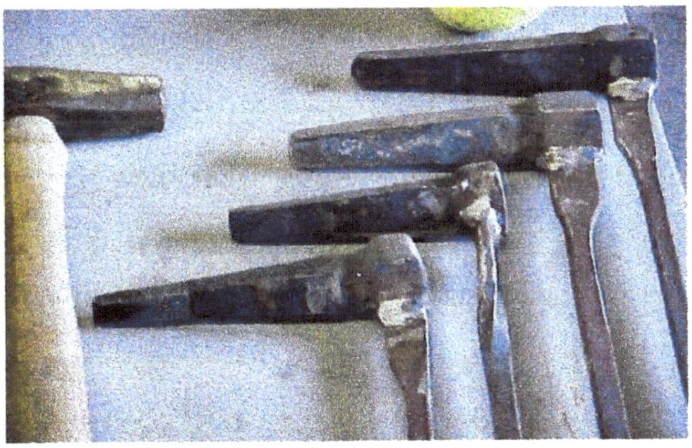

Picture 62

For these eye punches the handles were arc welded onto the working heads. Our preference here was 312 stainless rod (5/32) for attachment, a good selection for joining mild steel and tool steel. The heads (S7) were air hardened and tempered to purple before attaching our handle. Great care should be made to prevent overheating the welded union area as fracturing can occur when the punch is in use. All sharp edges on the striking surface are belt sanded off leaving the head crowned (highest in the center striking area).

Picture 63

When preparing for a production run the punches are often laid out as shown. Punches needing dressing (belt sander) or waxing (for lubrication) are made ready as the gas forge is warming up. The most appropriate punches for the project are singled out, checked for readiness, then brought to the work station for ease of use access.

Picture 64

A variety of set tools left to right: blacksmith's fuller for 1-1/2 inch radius, oval eye drift, crescent moon necking fuller, rivet head dresser, tong jaw top swedge.

The necking fuller is most often used in our shop with an identical bottom fuller hardy hole (square anvil hole) mounted.

Picture 65

Two tools at left are wood handled slitters. Some of our steel is slit before a hole is opened. Whether the final hole will be round, oval or square in shape, the slitters can help preserve material around a section. This preservation of material can be for structural strength or as part of a decorative design constraint. The four steel handled tools to the right are also slitters with offset heads. For the smiths that work alone, these are held in the tong side hand. This offset enables the smith to stand above the work with an unobstructed view of the progress. Many times the arms can also function more comfortably with this design. The handle covers shown are but 1 foot PVC pipe sections driven onto the distal portion. Anything to take some of the sting out of its use to the tong hand.

THE FORGED ROUNDING HAMMER STORY BOARD

Picture 66 (shown opposite page)

Shown is a story board for forging a 1-3/4 pound rounding hammer. These steps are the same whether the forging is done by hand or under a power hammer. The progression of steps is from top left down, then top right and down. Our finished heat treated, ground, polished and hickory handled end product is shown at the bottom. The taped billets are holding the eye buttons which have been pushed out. The student can then examine them for future reference. The fuller cuts to isolate the hammer head segments can be done before the eye is punched through, though not carried to their final depth.

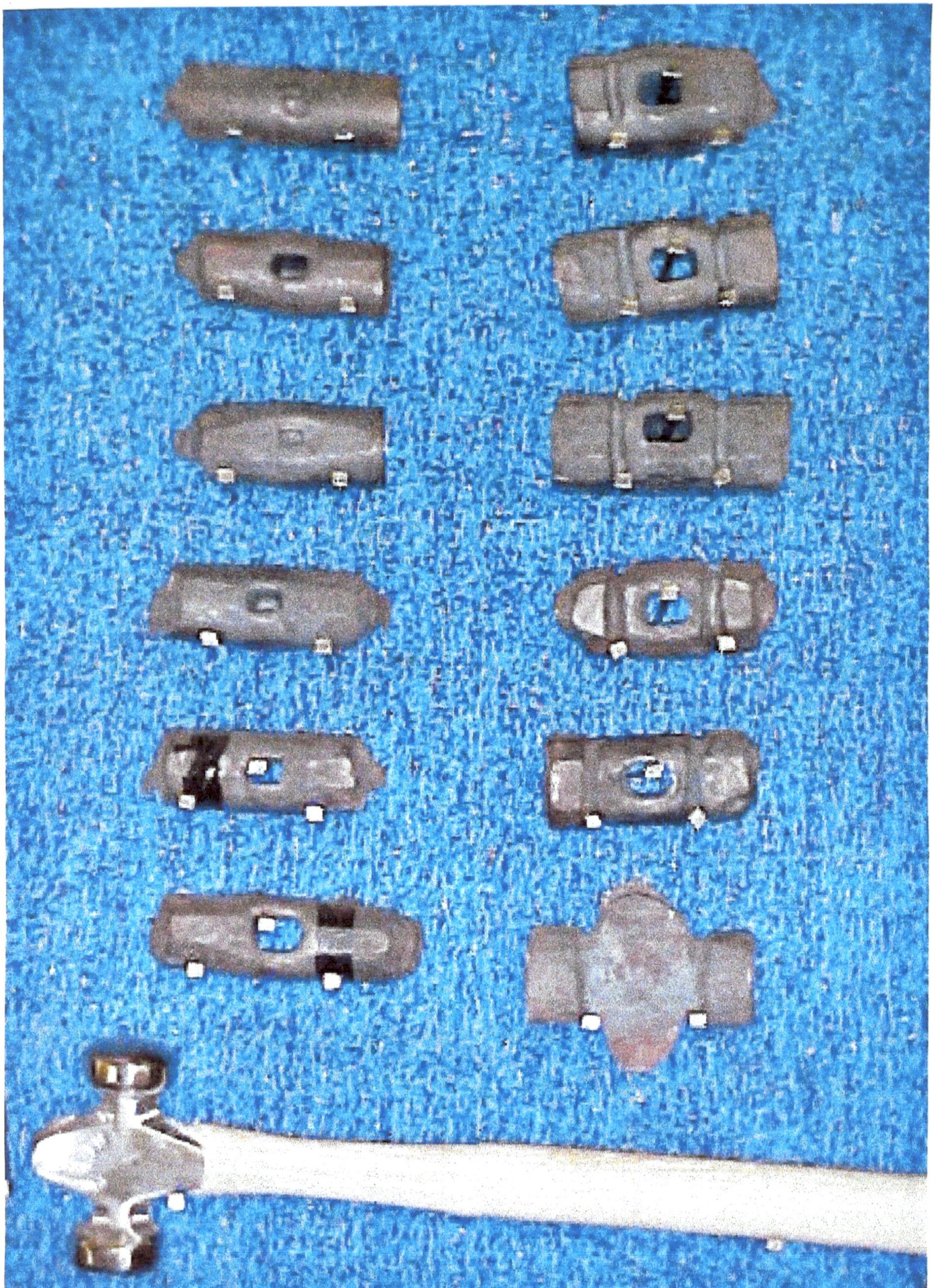

Picture 67

The hammer head steel is marked with an oval eye punch and checked visually for centering both in longitude and latitude. Flattening a center line along the full length of round stock will help keep it from rolling as this marking is done. The oval eye punch is then driven 3/4 of the way through this same mark to begin our handle eye hole as shown at bottom. The smith may elect to make all fuller cuts around this steel billet to isolate the striking surfaces from what will become the eye. This can be done with great results before any eye punching is commenced. For most handled tools with through eyes, the hole is punched first. The exception to this rule are tools such as flatters and tools with a wider working surface than the eye body.

Picture 68

Center billet is now turned over and a halo or cool spot is looked for. This spot will identify exactly where our eye punch started the hole. Line up the punch on this halo and again mark your billet as shown. The bottom center picture shows how deep the punch is driven and the center slug is about to be pushed out. This shows that the eye punch is doing its job and shearing all sides of the slug uniformly. Our eye punches with flat/blunt working tips will do this in a handy fashion.

Picture 69

Bottom center shows the eye slug pushed completely out and taped to the billet for reference at this stage.

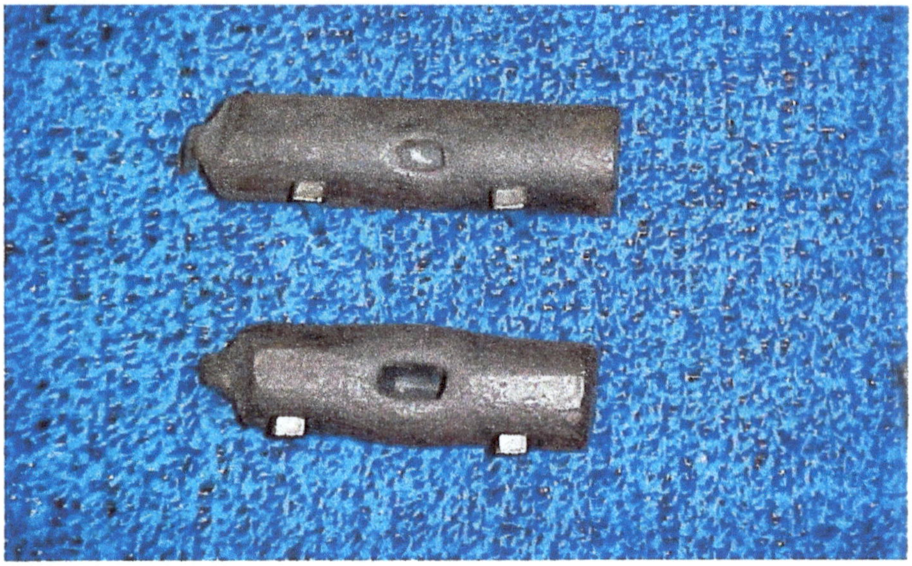

Picture 67

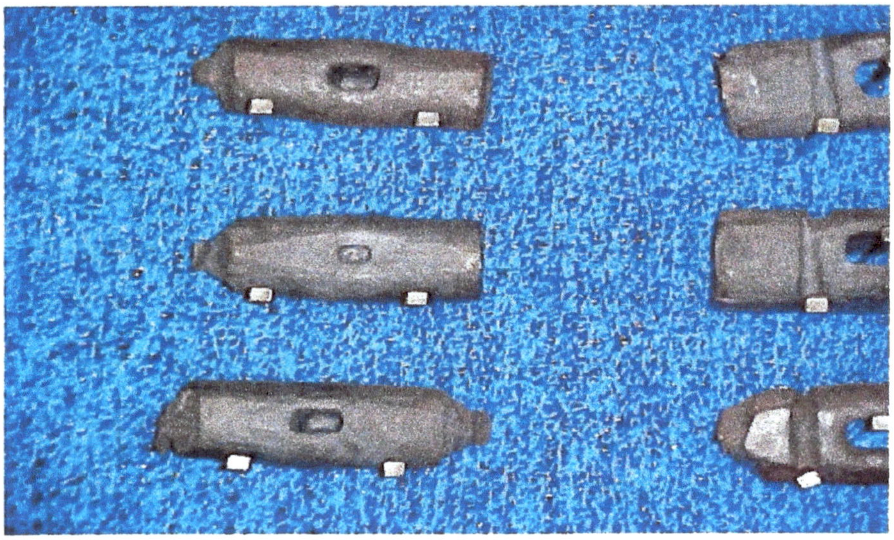

Picture 68

Picture 69

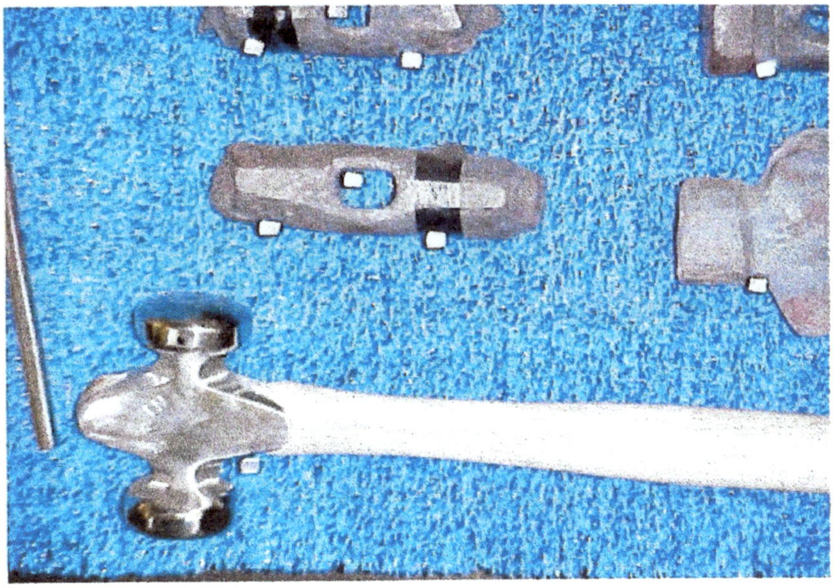

Picture 70

Center billet is showing the hammer eye beginning the drifting process, enlarging the eye with larger and longer tapered punch than our original. The flat side oval eye hole shape is emerging.

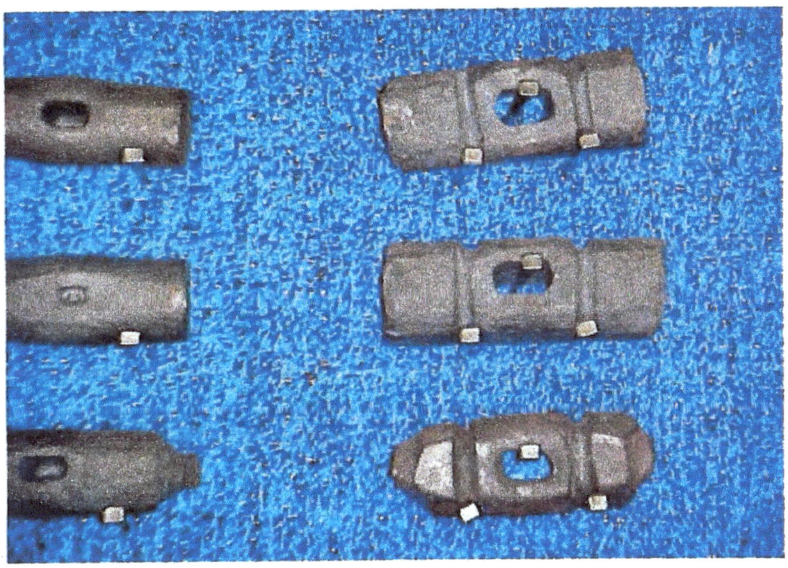

Picture 71

Top right shows fuller cuts made around hammer ends to isolate all portions. This fuller cut is made best with both top and bottom fullers working simultaneously. Rotate the billet to create first a square fuller cut. Next, to knock the internal fuller radius corners off to create octagonal. Lastly, rotate the billet to make the internal (concave) radius round and smooth. Circumventing the entire billet will give the smith a good look at how much volume has been pinched away for each striking surface. Bottom right shows both fuller cuts completed. We are now ready for eye elongation forging with a drift in the eye as a mandrel.

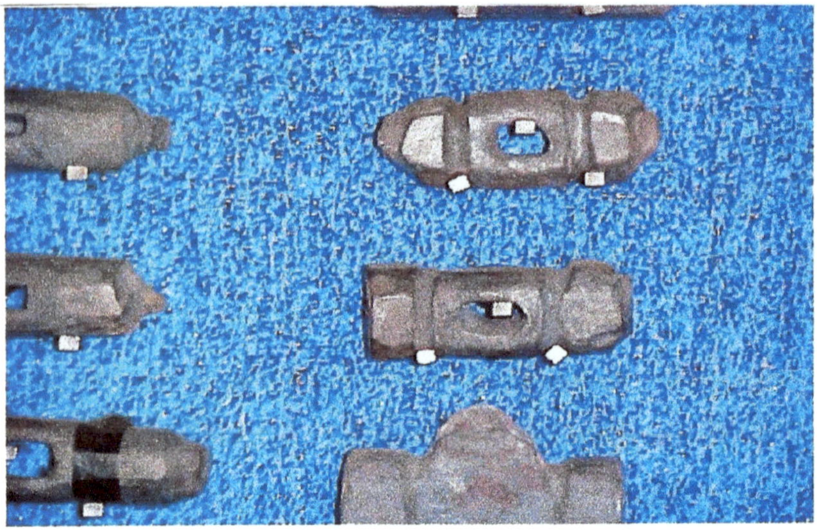

Picture 72
Making a storyboard of a project can be of tremendous benefit for the smith. Whether a project that is completed only once a year or part of the daily routine, having the actual steps from one stage to another can save countless hours of rebuttal and despair. All of the forging on the hammer heads shown was done without the use of machinery.

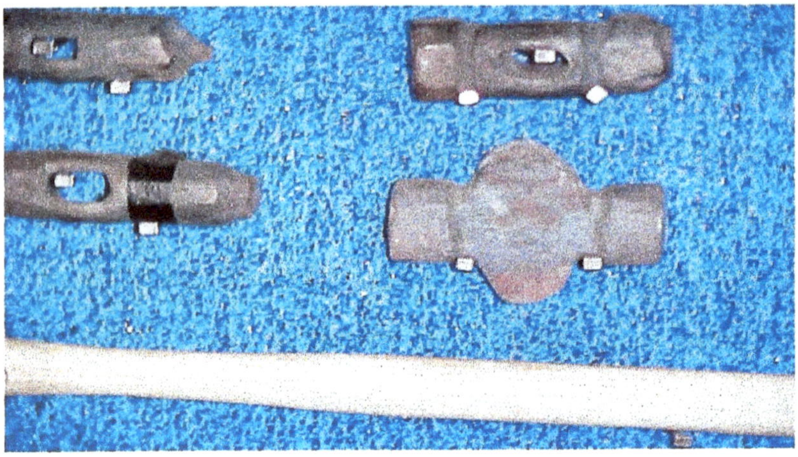

Picture 73
By elongating the eye more mass will travel in a linear fashion. As a swinging tool there is much less lateral deflection after impact when in use with this design. The bottom right rounding hammer is finished in terms of forging only. After this last forging heat, it is reheated to a cherry color and placed in a large wood ash can for annealing (slow cooling process).

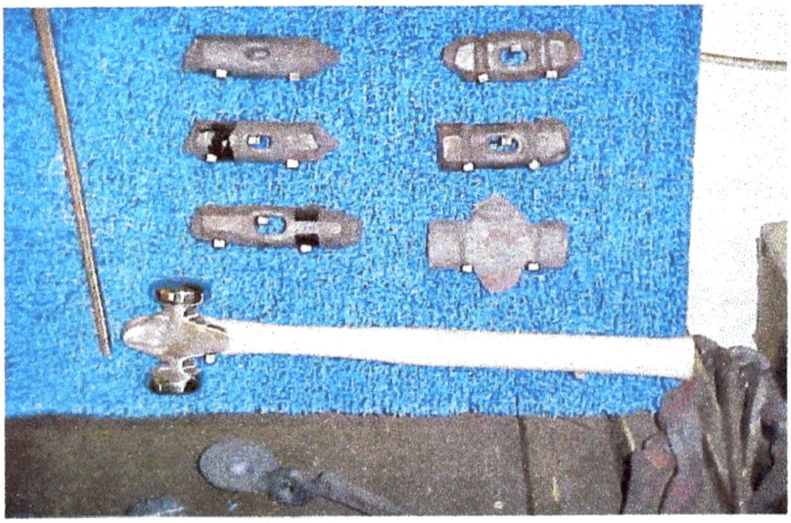

Picture 74
The bottom rounding hammer shows the finished product achieved through the consecutive progression of the previous steps. Many smiths can remember the steps needed to complete a project, but over time the sequence can become fuzzy. Having a board such as this for some of your favorite forging projects will help write it in stone for posterity.

TONG ALIGNMENT: JAWS AND REINS

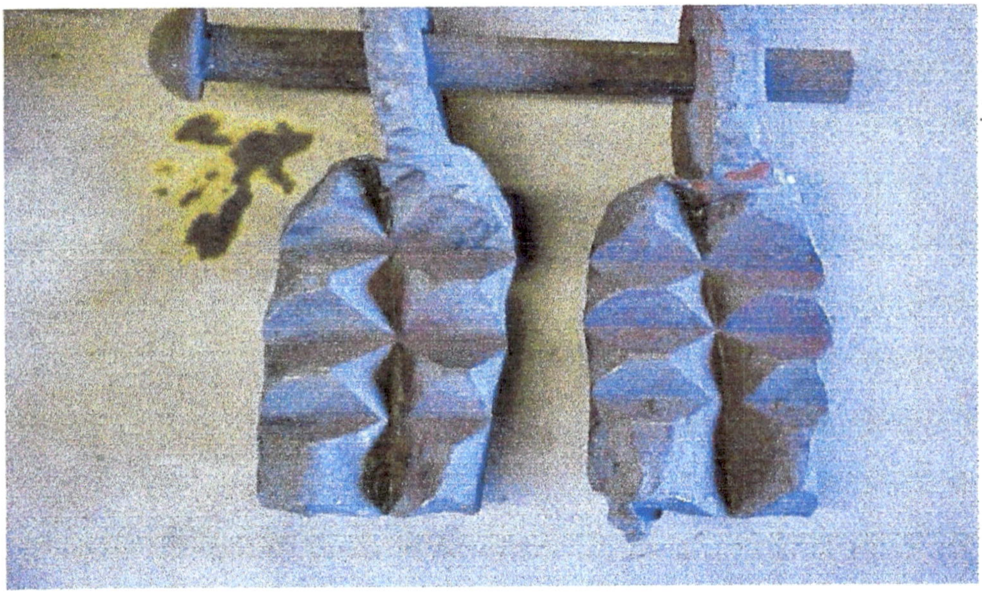

Picture 75

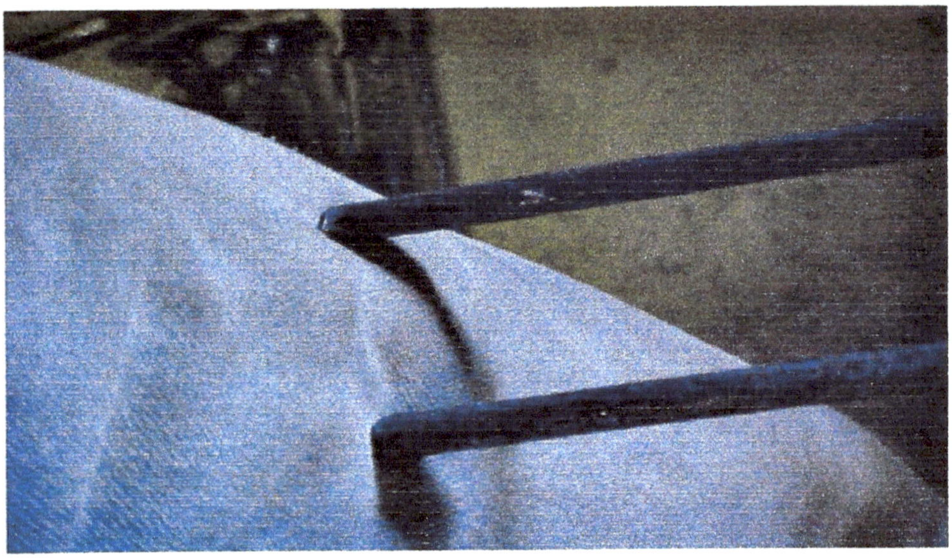

Picture 76

Pictures 77

Picture 75 (opposite page)
Hot marking is made singularly from the jaw lips to meet with the bearing surface edge of each jaw. They do not need to be connected with their mate as shown here but this view enables the reader to see what they both must look like. This was done with a 90° convex V top tool. The cut described here is for holding material lengthwise, all other cuts shown here are called crosswise.

Picture 76 (opposite page)
Quenched out cold rein ends can rest against the smith's leg when working alone. Hot jaws of tongs are on the anvil flat face.

Picture 77 (opposite page)
Wood handled hot chisel used to mark each jaw simultaneously. Groove alignment will now be certain. Marks do not have to be deep, as a cold shunt can result when the final depth is made with the V cut tool.

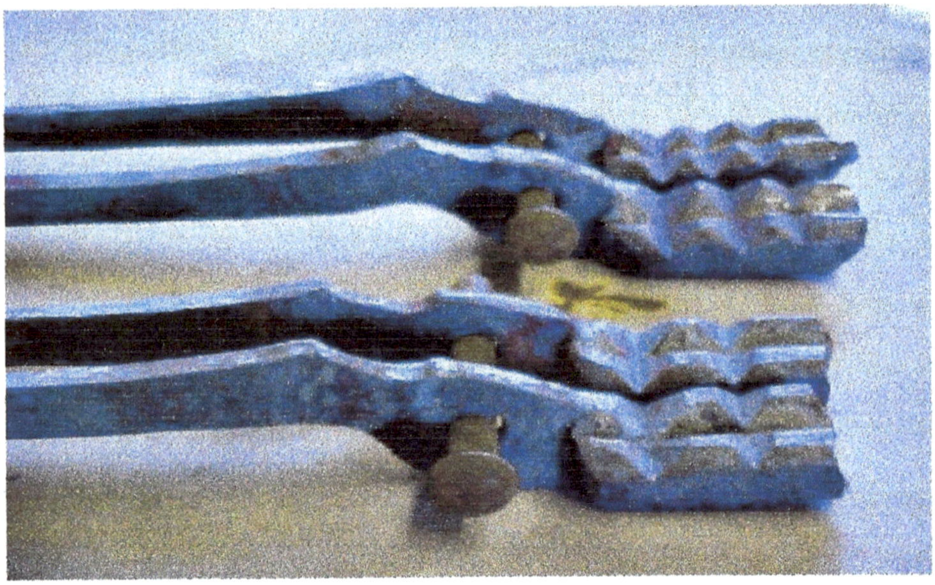

Picture 78

Wolf jaw tong (hand forged). The rivet holes are pinned before the jaws are grooved.

Picture 79

Tong blanks are run off in pairs. Their mate is aligned with a long rivet of the correct diameter for that rivet hole. They are next placed side by side as shown after heating and retrieving from the forge.

Picture 80 (see opposite page)
Box jaw combination flat jaw tong can be set up many ways. The flat cross hatched side here was made to equal the outer width of our boxed uprights. Fitting the flat jaw to pass internally to the box is also regularly made.

Picture 81 (see opposite page)
At left a crisscrossed V cut flat jaw tong ready for finish grinding and assembly. These were marked for grooves separately, then a long rivet slid through both halves to check for similarity. The truly hard made way. A good suggestion for this type of grooving is to center punch both blanks when they are side by side. Next make sure that all deepened grooves or cuts bisect **that** center punch mark.

At right, a side jaw tong intended to hold 1/8 inch flats. These are hand made to be pairs from mid jaw downward to the distal (far end) of the reins. From mid-jaw to the tips of the lips all adjustments in forging are made one jaw at a time. All to insure that our inside radius of the outer jaw matches the outside radius of the inner jaw. A rivet can be dropped through both halves as pictured. A piece of steel - they are intended to hold - then placed in the jaws to check for gaps or imperfections. Corrections in forging are then made one jaw at a time!

Picture 82 (see opposite page)
A pair of triple bite pick up tong. The rivet hole again used to keep each curve identical in width and depth to its mate, the opposing tong blank. These were done one curve at a time starting nearest to the rivet hole. A one inch radius convex top tool and a one inch concave bottom tool for the first curved depression. They were curved side by side as pictured with the rivet inserted. Consecutive curves toward the jaw tips were made with the same top and bottom tool combination but not driven as deep into the jaws. Tooling can be changed as the smith's desires dictate.

A die can be made to forge all curves at once for both blanks. This may require the use of your power hammer, hydraulic press or a big helper with a sledge hammer. Any combination of "V" cut to hold square stock, half round or hold round (as shown) and others are possible using these methods regardless of whether under power or singular curved by hand. The heft and length of the jaws and reins are dictated by the work they will hold. When making a pair of these pick up tongs simply on the horn of your anvil or with a cross pein hand hammer, make a decision. To either hammer into the valleys or hammer out the hills. Then do the same for both jaws to make a pair.

Picture 80

Picture 81

Picture 82

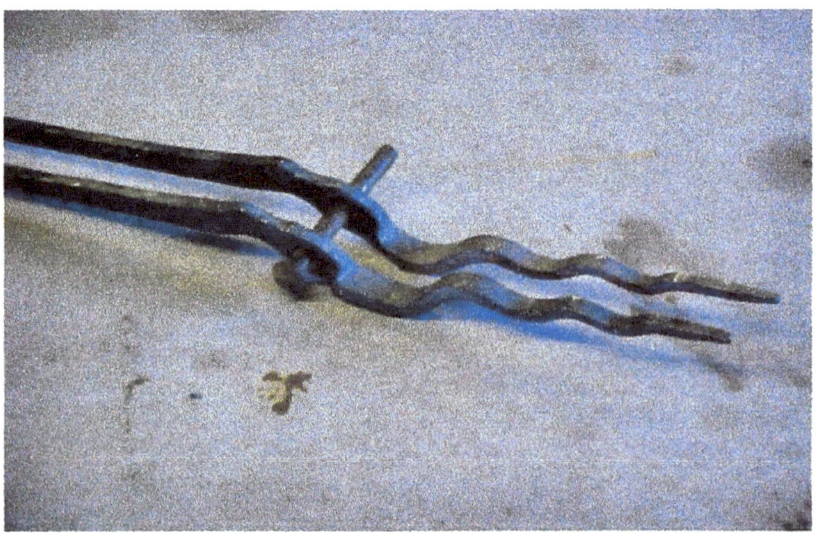

Picture 83

A specialty use tong such as these flat jaw 3/8 round cross bits are made identical by first punching the rivet hole after the blanks are forged. The tong shown was hot punched as they were made from 1 inch jack hammer steel. It's very hard to drill a rivet hole through this steel unless annealed, but drilling is fine. The main consideration is to approximate each half of the tong blank similarly. The rivet hole is then used as your benchmark to make the jaw fuller grooves identical. Each tong is heated with its mate, they are drawn from the fire and a long rivet inserted through both halves as shown; its then rested on the flat face of your anvil. Fuller marks perpendicular to the jaws are then made simultaneously. They now have no choice but to oppose each other when the tong is assembled. The fuller grooves parallel to the jaws are done one blank at a time. These marks merely meet with the bearing surface side of the tong boss (where one blank touches the other).

Picture 84

Tong blanks which are not identical can also be made to pair up when the rivet hole is used for their alignment gauge. For these box jaw combination cross hatched flat jaws, we prefer the box side to be slightly longer. The extra length helps in retrieving projects from the fire when in use.

Picture 85
Jaw lip view of tong left to right “V” jaw hollow bit, triple bite pick up, chain makers/rivet, alligator/needle nose.

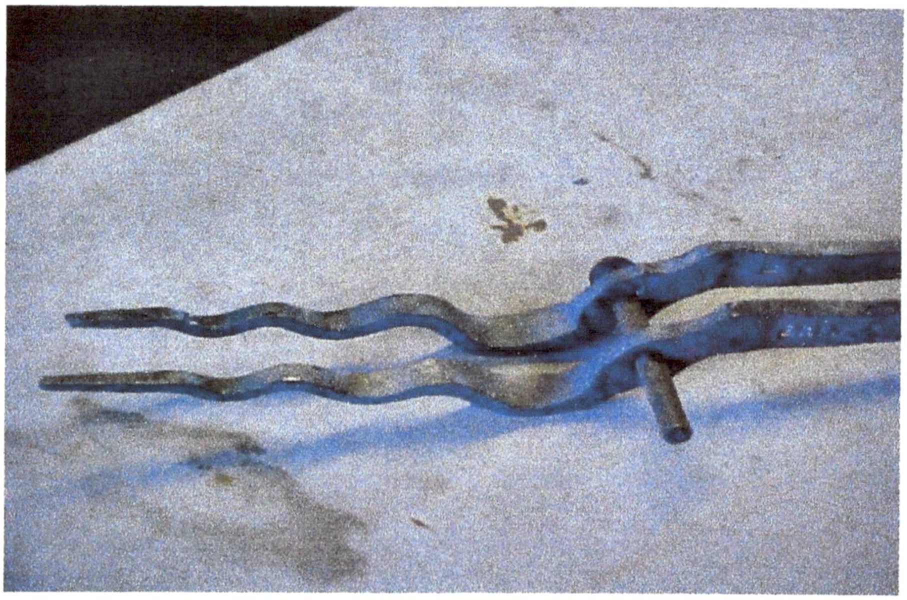

Picture 86
Unfinished triple pick up tong the same pair as shown finished below.

Picture 87

Hollow bit tong blanks are checked for similarity. They will become the V jaw hollow bits shown below right finished. Note, the near ground rivet hole has been cross hatched with a jack hammer star while hot. This will be the rivet lock side. The proper length rivet will be upset into the divots as pictured.

Picture 88

Left to right: alligator pick up tong, great for fire brazing. A chain making tong; the end notch holds 5/16 inch round stock. Finished triple pick up tong done with a cross pein hammer into the valleys off of our anvil horn. A "V" hollow bit tong sized to hold one inch hexagonal stock. All were made of 1 inch diameter hex jack hammer bit steel, 11 inches total per pair of tongs.

Picture 89
A very wide (3-4 inch) straight chisel (non-handled) can also be used to mark your jaw grooves. If half round grooves are desired, this marking should be done with a blunt chisel. This will make for a cleaner end forging without defects.

Picture 90
The magic "V" top tool used to drive the grooves to their final depth. Wooden handled, this tool shown is made of S7 1-1/4 inch square, hardened and tempered to straw.

Picture 91
The "V" tool displaces and contours the grooves, so it is important to make sure that opposing marks are driven as closely as possible to the same depth—especially critical in multiple grooved tong jaws.

Picture 92
This V tool handle was attached with a bicycle tube section slid over the through wood. The excess tube is trimmed away to prevent ambient heat ignition.

Note the tong jaw at right showing the graduated depths. Its opposing mate is made identical. One pair of tongs can now hold many different diameters of material.

Picture 93
At left, blunt edged marking chisel used to mark 1/2 round concave grooves. At right, sharp edged hot marking chisel used to mark for grooves to be deepened by the "V" top tool.

Picture 94
A finished pair of wolf jaw tongs. All grooves were made with the same top tool shown above.

Picture 95 (see opposite page)
Side view of a quadruple bite wolf jaw tong. One size of material round or square can be held in the jaw lips. Four different diameters of round or square can be held perpendicular to the jaws. All in one pair of tongs. . .

Picture 96 (see opposite page)
The tong alignment gauges are made of all different thicknesses of steel to adjust each individual tong. Quite often the hand forged tong jaws will not center the gauge between the reins when squeezed. This gauge will tell the smith where the problem is. Occasionally we may want the jaws to point offset from the reins for a specific grasp or project, but even still, the jaws must do this uniformly and together with its mate for good holding security. This gauge is the acid test for tong alignment. Straight lipped tongs will point the gauge across the rivet and bisect evenly the space between our reins.

Both jaws should be bearing all their internal (working surfaces) on the gauge as the vise is tightened to hold.

Picture 97 (see opposite page)
Opposite side view of the tong gauge. A tong being adjusted for final fit should be looked at from many sides and above to find your best reference point. Both reins or jaws may not be off from ideal—sometimes only one half is the culprit (the jaw of one or the rein of the other).

The 90° turn in the gauge was bent while the gauge bar stock was in a vise to maintain exact dimensions of all portions held between the tong jaws. Remember that as steel is bent or forged in a turn, the inside radius will thicken and the outside radius will thin. This factor of steel dynamics should be corrected if excessive to enable the gauge to operate correctly.

Picture 95

Picture 96

Picture 97

Picture 98

Shown is a pair of flat jaw tongs being sized for holding 5/16 inch thick stock. The gauge inserted between the tong jaws aids in achieving the proper jaw width and rein width for the smith's hands. This width is the distance between the reins; which will vary from one smith to the next. As we all have different hand sizes.

The tong alignment tool

- used for hand or power hammer forging tong pairs
- assures equal stock placement between jaws to reins
- Used for most all sizes when equidistance <u>is</u> desired of reins
- separate alignment tool made for each stock size

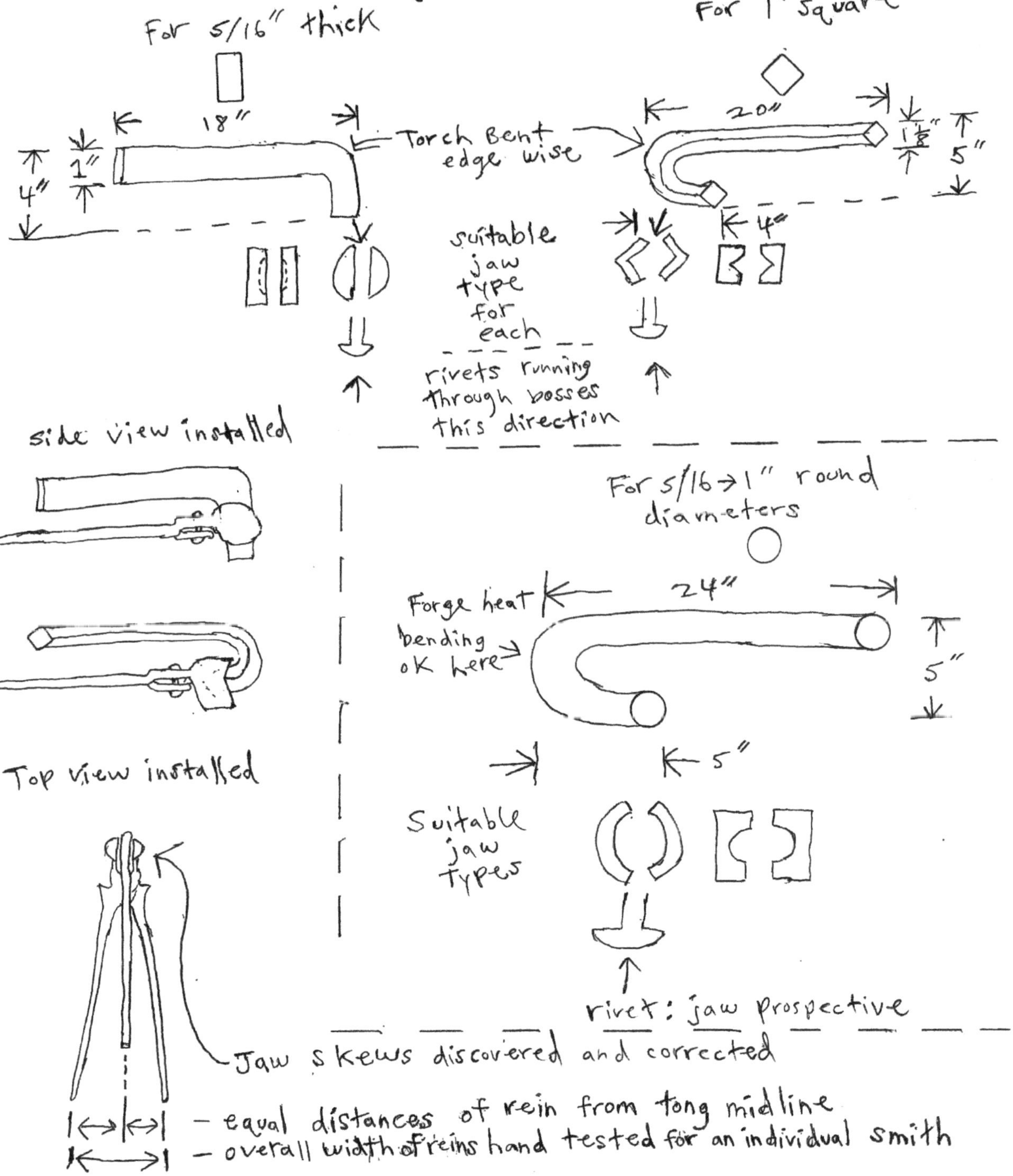

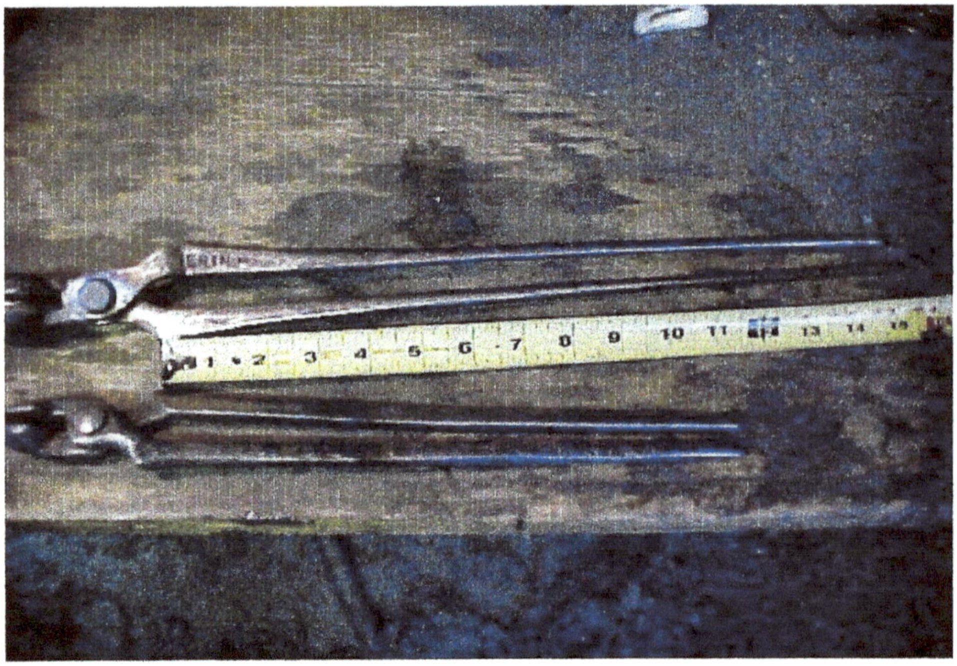

Picture 99

One of our chief pet pieves is watching a smith, new to the profession, trying to work diligently with a tong that has very short reins. When making a new tong by hand, much of the forging effort may go into drawing reins to a decent length. Many smith's fade here unnecessarily, thereby must put up with excess heat to their hands and body. We are describing heat proximity considerations.

Whenever that tong is in use, discomfort is inevitable. Jump welding but a few inches of round stock onto these short tongs will make a world of difference. Review an old pair which has lay idol for a while to see if that has been the problem. Many reins can be added to without extracting the rivet. Forging the reins to greater length is also excellent, provided too much strength is not lost.

The rein length considered ideal for our light work (e.g. horseshoes) in coal or gas fired forges is 15 inches. That is the length measured from the distal shoulders to proximal end. Rein diameter is dependent upon the raw materials used for their forging. The pair above is made of titanium (low magnesium) thereby enabling a thicker and wider rein without excessive weight. With most all materials there is an ideal ratio of size to weight to strength of holding power. They must feel comfortable in our grasp as well, for security develops from this ratio so now all can search for a little....

Forging on the far side of your anvil with a tong that has added rein mass for length is helpful. To the near side anvil forging sometimes may seem as a hindrance; they seem heavier because of mind air suspension, nothing to support the forging but muscle and sinew. There is more counterbalance weight to forge against. Tendering that tong from forge fire to anvil or power machinery with a billet in tow may seem heavier, but how does it perform in use. Weigh the benefits in advantages versus disadvantages.

Picture 100

An old tong can have a new lease on life if rivet replacement is made properly. The old rivet head was ground off. The remaining piece hammered out with a round punch. Inspection for rivet lock cuts is made. One boss is heated and lock re-cut for proper depth then air cooled. Both holes re-drilled on our drill press for size and round trueness. A new rivet is installed hot and peined over. We are looking at the rivet exit side with lock cuts exposed at left.

THE CHAIN MAKERS TONG

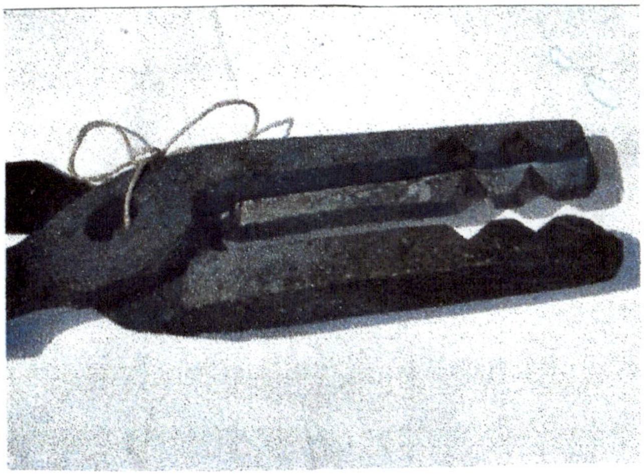

Picture 101

The "Chain Maker"

Here lies a pair of chain-makers tong blanks. Forged, paired up and tied to its mate. These will be stored on a shop shelf until such time it is necessary to make chains from 3/8" round stock. The jaw cuts were all made with our "V" cut handled top tool to oppose each other.

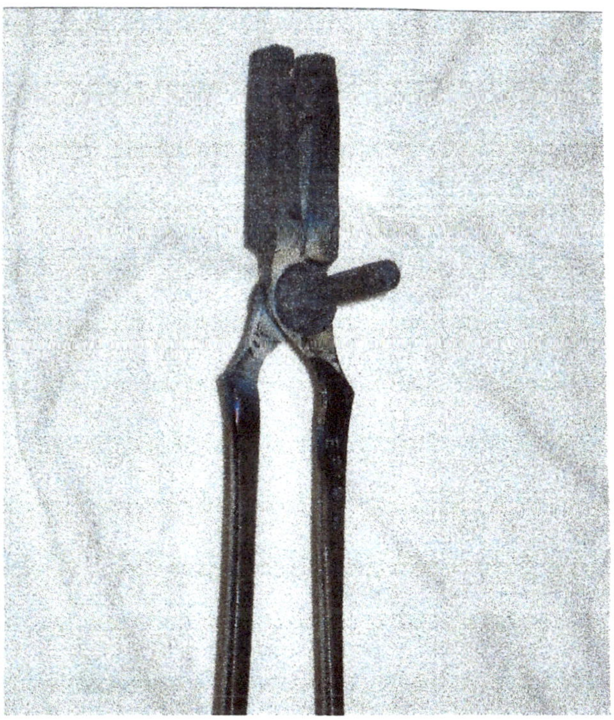

Picture 102

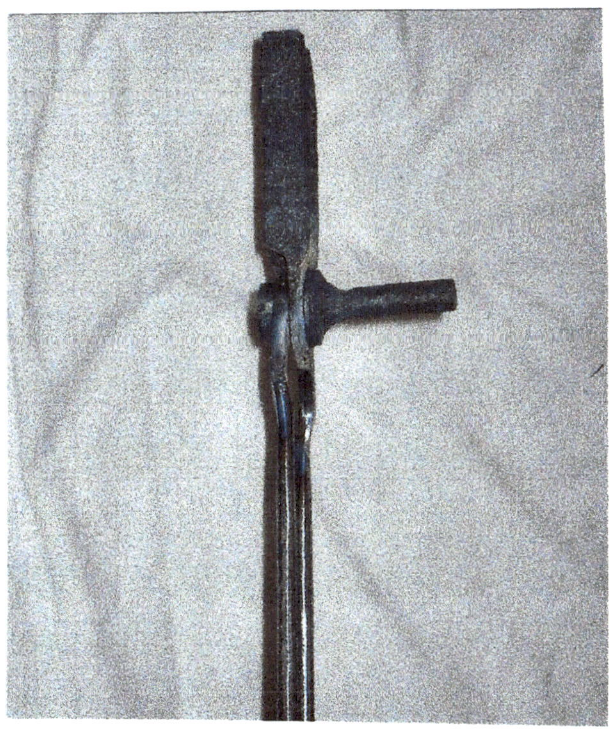

Picture 103

The Chain makers Tong Rivet

materials : A36 (mild steel)
round 3/8" for 5/16" Tong blank hole
1/2" for 3/8" "
5/8" for ½" "

eg. ½" × 2½"

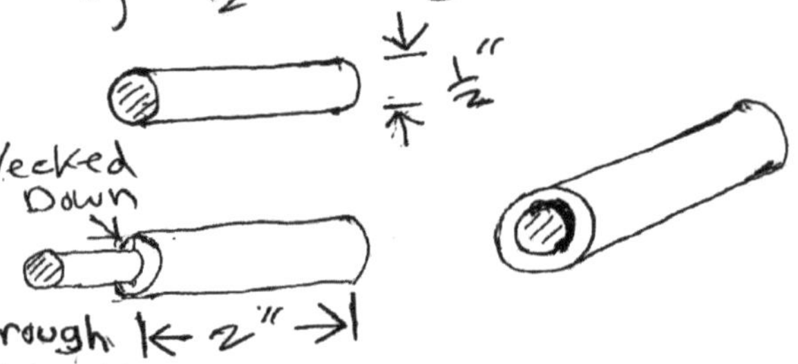

will go through both Tong blank halves plus 1½ times the diameter protruding

Top View

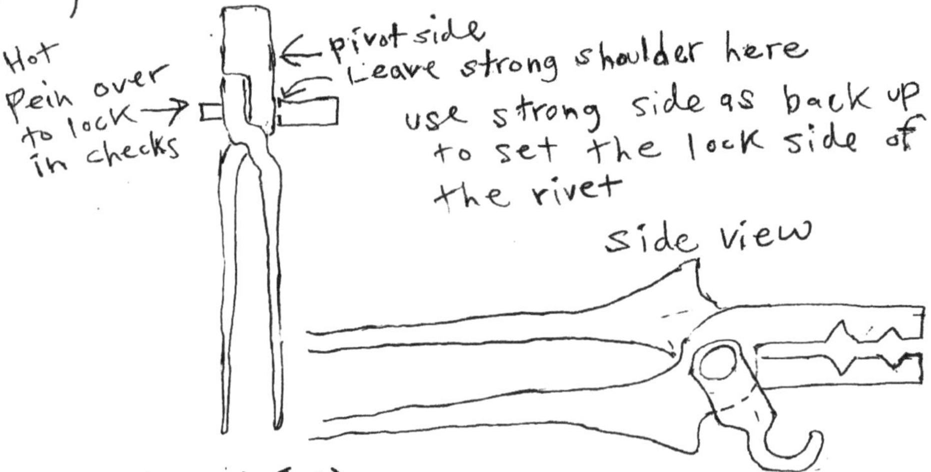

- Draw length (1") to rivet strong side after installation and hook it.
- Hook tip point to the sky when finished and in use.

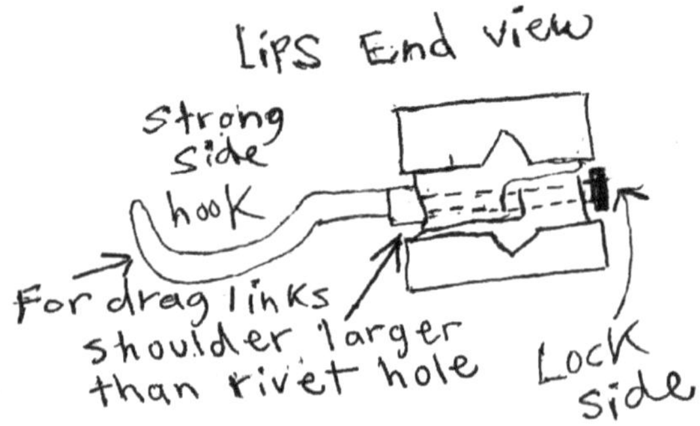

THE HEAVY TONG FOR ANVIL MAKING

Picture 104

Left to right: "V" jaw tong for holding 4 inch round. Used to forge large wrist pins into anvil tops. Made of mild steel with twisted jaw shoulders.

A "V" jaw tong twisted 120° with lock ring on reins. Used to hold the heel or tail of anvil tops when forging the horn; made of a jack hammer bit; air cooled after forging the tong.

A box jaw tong for holding that portion of knives not being forged. Grip range is 3/16 – 1/4 inch x 1 inch comfortably. Made of sucker rod from 1 inch round. Note the size differential as it is dwarfed by the anvil tong.

With most any tong, it must be made with enough heft (mass of jaws and boss) to accept the forged billet heat without fatigue!

Picture 105
At left is a large 4 inch round track pin from a tractor. It was forged in the position shown with the keeper pin hole sideways into the anvil top beside. These were both from identical pieces of steel.

Picture 106
No edges have been ground yet, only crisp from forging.
The makers name is hot stamped in for posterity.

Picture 107
In forging larger pieces of steel, the added volume can really show itself. It seems to grow as the two pieces were identical before the forging began. Tong heft and length should also grow proportionally. At some point a jib crane or overhead trolley will become necessary to handle tool and billet weights steadily.

Picture 108
Horn tip view of forged anvil top from our track pin. A tennis ball is placed in the picture for size reference.

Picture 109
Anvils of all sizes can be made from a build up method. All parts shown were forged from wrist pin remnants of 2 inch round. Round and square hole steel handled punches were used to make the pritchel and hardy holes respectively.

Picture 110
The anvil is ready for heat treatment, arc welding for attachment, final grinding and polishing. The anvil can also be assembled by forge fire welding and later heat treating. Test yourself to see which method agrees with current skills. Start small and then build up to the bigger forgings. Many of the skills needed are similar but the scale expands. An optimum for size and intricacy will be discovered. A feel for maximum will unfold and new ideas which need to be implemented.

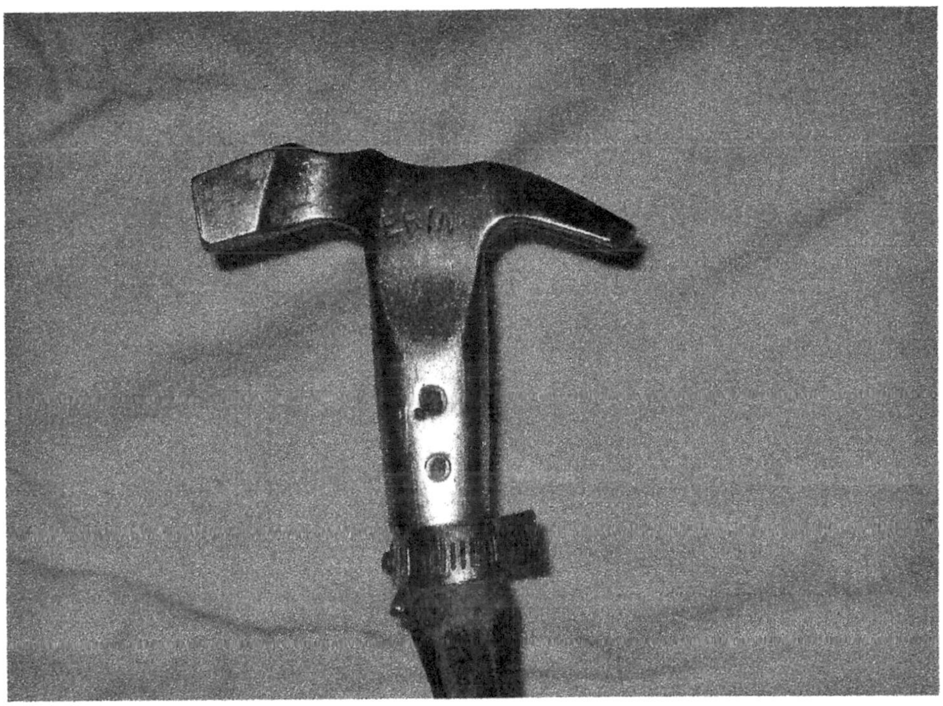

Picture 111
The blind hole type shown here.

Our actual Blind Hole claw Hammer Traced
Head Weight 16 ounce driving (farriers type)
10% wider & taller face than poll for construction type

'Nomenclature'

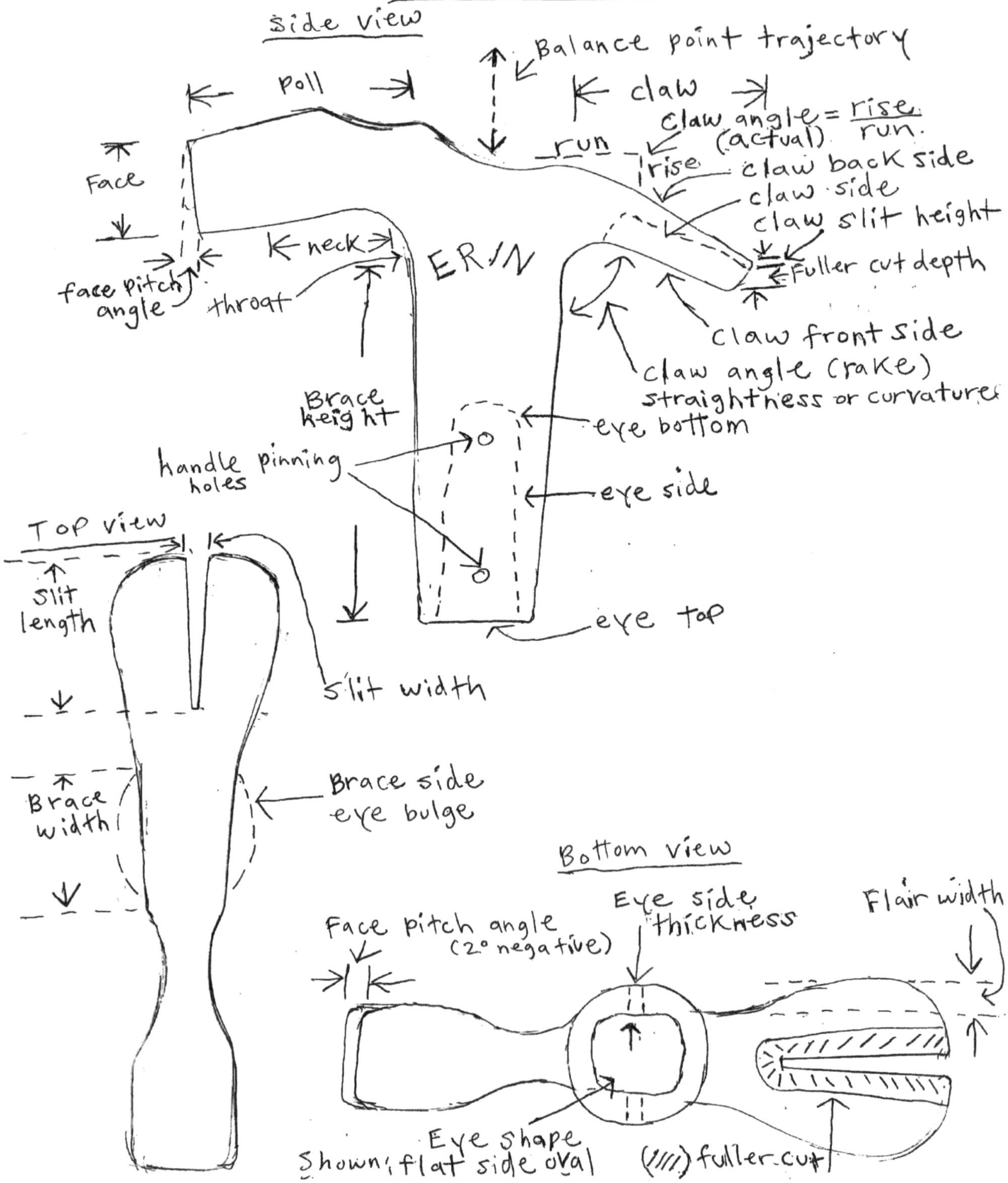

Driving hammer with blind through hole
hand forging progression

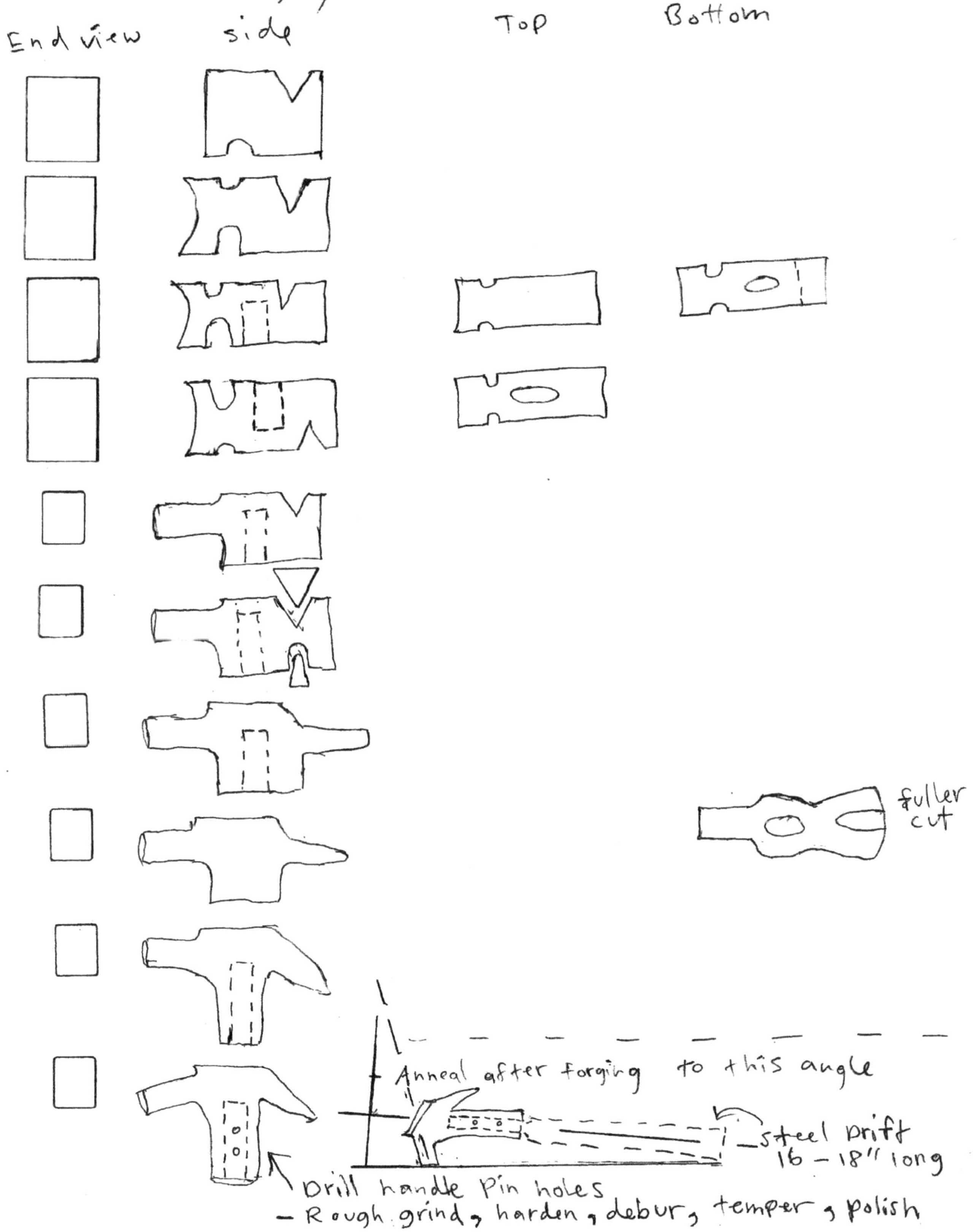

The Blind Eye Claw Hammer
Another way to get it done!

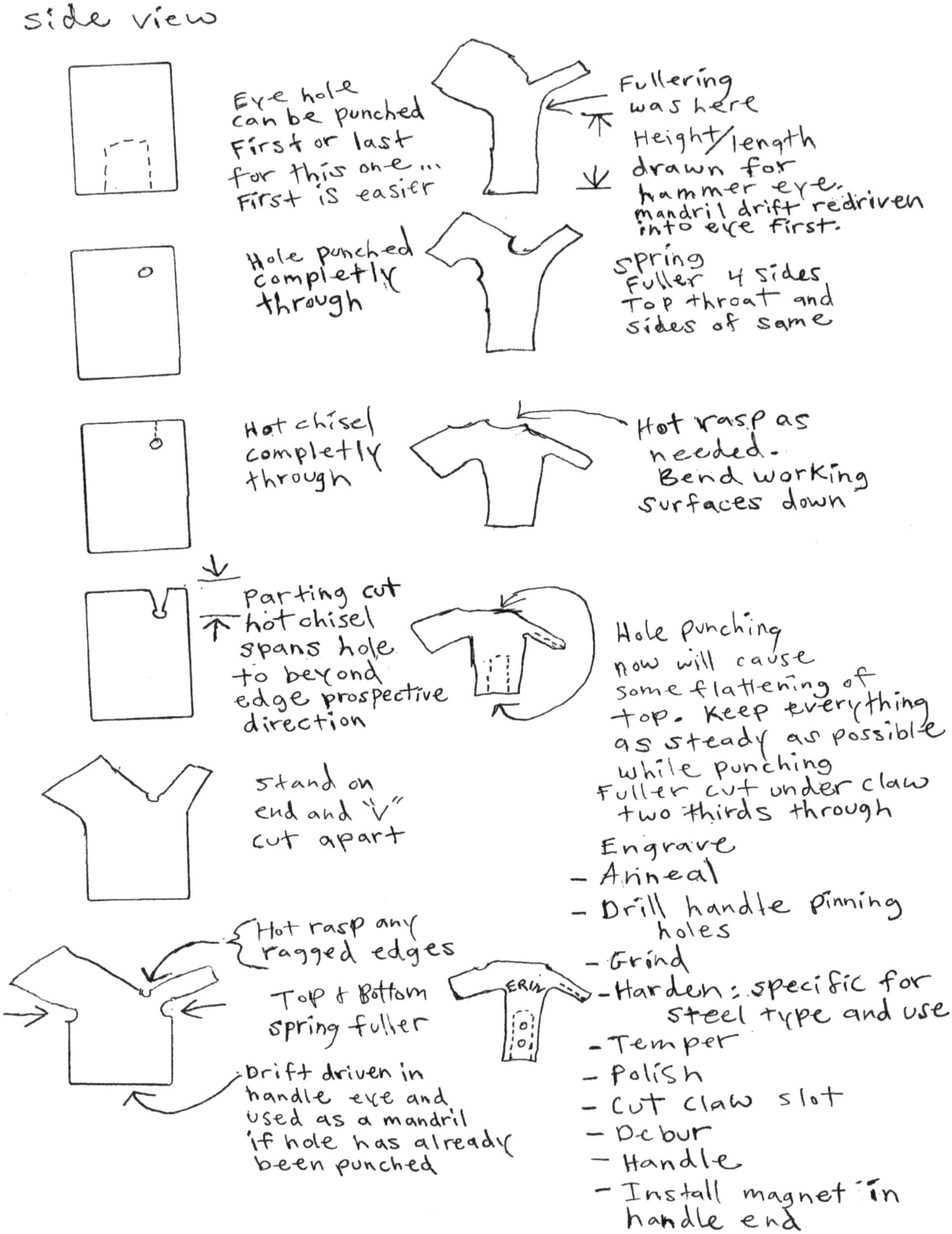

Forging Progression for 10–14 ounce claw hammer with through hole eye
$1\frac{1}{4}$" x 4" Hexagonal Jack hammer bit remnant

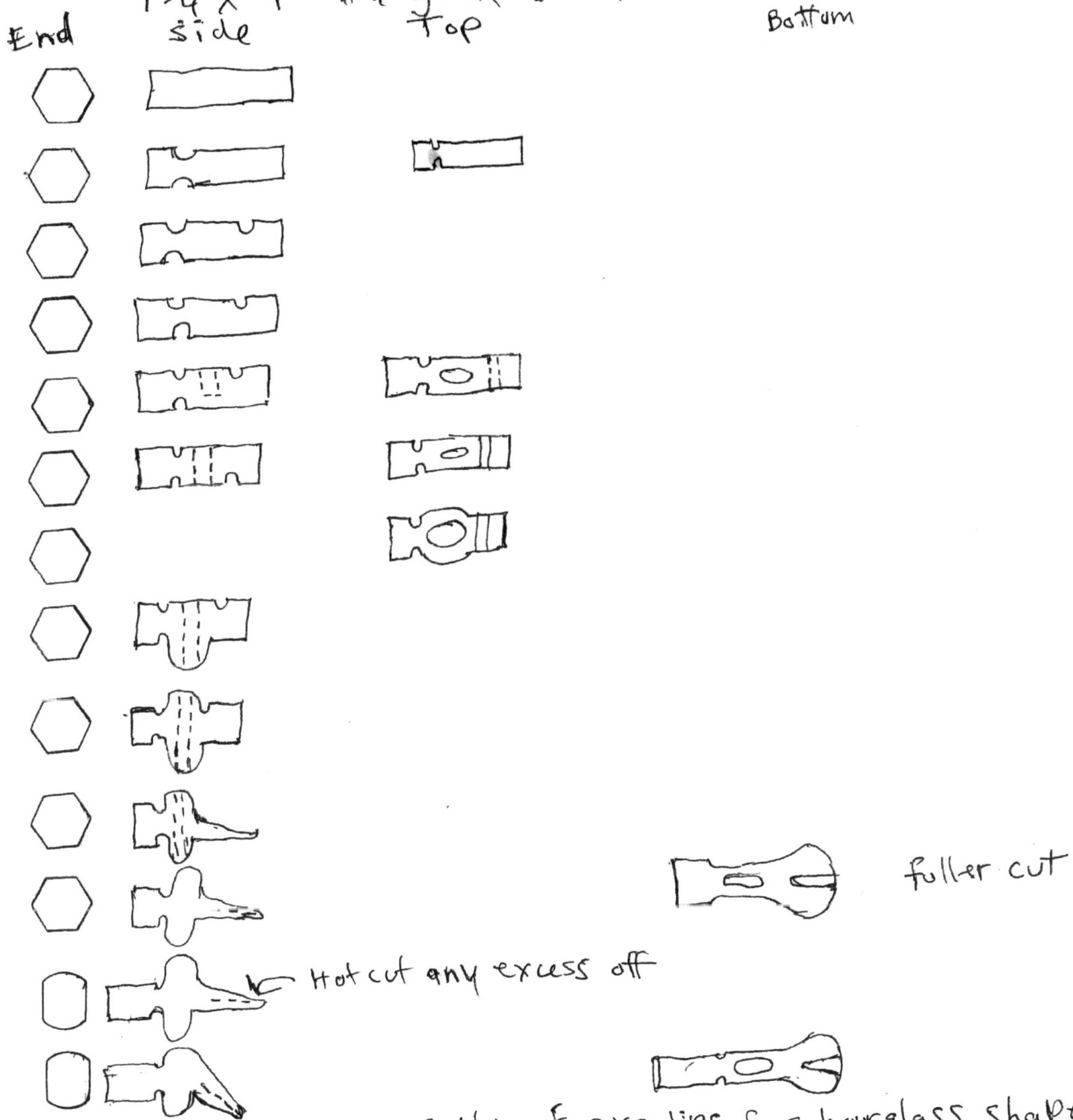

part top & Bottom of eye lips for hourglass shape
Engrave
Anneal
Rough grind
Harden (water), debur, temper (Blue eye, straw face & claw)
Split claw
Polish, handle

The following segment, when held up to indirect sunlight, brings the reader right into my shop.

Picture 112
Using the set top and bottom fuller to isolate material for a claw hammer, the dimensions of material is 1-1/4 inch diameter hexagonal; cut to 4-1/2 inches long. This was a remnant jack hammer bit; solid with no internal hole.

The fuller is isolating material from the striking surface and the handle eye area.

Picture 113
Here our claw hammer is resting on the trip hammer bottom die held by tong shown left. To the right a top fuller is being driven into what will become the upper claw portion of our hammer. This fuller cut will isolate material as well as start the claw forging for this style of hammer.

Picture 114
Now that the working ends of the small claw hammer have been isolated, the hammer is ready to have its handle hole punched through. This particular bottom die has a 1-3/8 inch diameter round hole all the way through at right. When a hole is finished punching for a tool head, the punch tip has a place to go.

Picture 115

The oval eye punch is moved into position and upper die is engaged to deliver a blow. The eye location is first marked and checked to ensure good positioning has been achieved. For us, punches and set tools are held in the right hand, tong is held with the left. When the smith is happy with the eye mark location, the punch is driven 2/3 of the way through the billet. The sides and ends of this billet are hammered lightly and the eye punch can be extracted from your newly formed hole.

Picture 116

Another heat is taken and the claw hammer billet is placed on the bottom die blind side of the hole upward. The oval eye punch is again placed in position and the upper die can begin delivering a blow. After two or three hits, the punch and billet are moved from the flat bottom die section over to the right where our hole is waiting. The eye punch can now push out the eye slug (button) and clear the billet eye out. Punching is finished and now we are drifting.

Picture 117 (see opposite page)

The oval eye punch has been driven completely through our claw hammer to be. The punch and attached forging are laid on their side to the right near corner of the bottom die. Two taps from the upper die and the punch with forging still in place is turned over and rested on our left rear corner. Again, two taps from the upper die are imparted. These taps are forging the eye sides to flat and we are now using the eye punch as a mandrel to prevent hole collapse. Note how the eye punch handle comes away from the working punch offset at approximately 25° away from in line with the punch. This is essential for several reasons: (1) to give the smith an unobstructed view of the forging, (2) to enable rotation of the forging to all parts for punch extraction, and (3) to create comfort for the smith as our hands and arms are not crowded together, holding a tong in one and top tool punch in the other. An offset punch handle here creates a more comfortable and efficient work space.

Picture 118 (see opposite page)

This is the third position used to place our forging for punch extraction. A tap or two from the upper die is all that is needed. The protruding punch tip is then bumped against a large anvil face and the claw hammer blank will slide off the oval eye punch. Any corrective punch hand hammering is now done to true up this important tool. It can be quickly done as the heat to accomplish truing is now born exactly where it is needed. This residual heat in the punch was picked up from our claw hammer blank.

Picture 119 (see opposite page)

Shown is a clean oval eye with flat sides, freshly punched and the punch tool just removed. Our “V” jaw hollow bit tong is holding what will be the striking end of the claw hammer. The material to the left of this punched eye will become the hammer claw. The single top fuller mark previously made is away from our view. We are looking at the under side of the claw.

All these steps, hole punching and tool extraction, take place in an approximately 20-30 second time-frame for a small (14-16 ounce) hammer. Though speed is not the essence of our focus, we use expedient means here to spare excessive heating of our punches. Each hole punch tool is rubbed with bees wax after it air cools to warm. This helps tremendously in the tool extraction process. Glycerin soap will also work well for this type of lubrication. Both products have been proven cost effective and are readily available. Numerous high tech lubricants are currently available if the smith will seek them out.

Picture 117

Picture 118

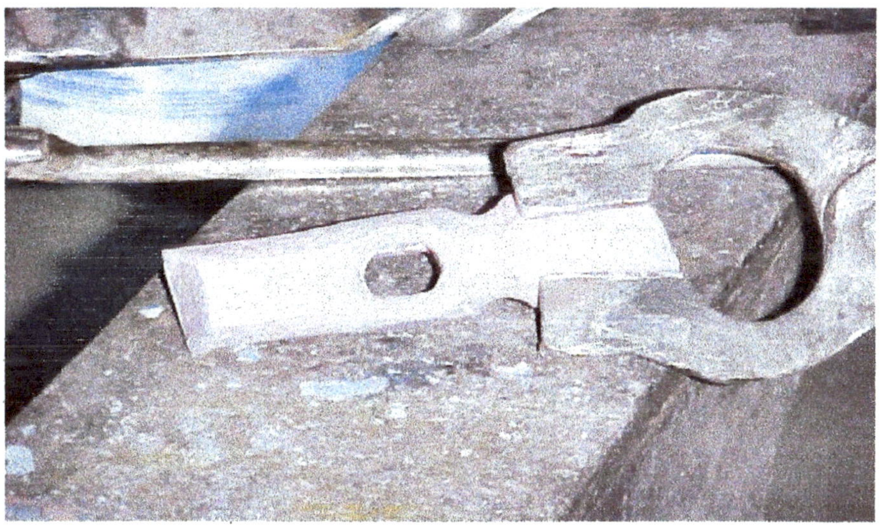

Picture 119

Picture 120

Trip hammer operation can be of great benefit for production work. Safety precautions should be ever present. Note the size and heft of the guard surrounding the moving parts in front of this 100 pound power hammer. It is made entirely of diamond plate steel one quarter inch thick to prevent the operator from serious injury.

The forging project at hand is using a top and bottom set fuller saddled to isolate portions of a farrier's driving hammer.

Picture 121

A close up view of a set fuller. Note the calibrations to the picture right side. They are in increments of half inches to help the operator find identical depths of fullering from one tool to the next. The upper die is in the down position here and finishing a blow.

Picture 122

The claw has been drawn with our flat dies to an approximate angle under the power hammer. It is now time to make a narrow fuller cut to split the claw lengthwise. The narrow fuller used for this does not need to span the entire length of the claw. A shorter one is actually better for walking the fuller along. This shorter form is not fuller height. It refers to length of the fuller working surface. The tool is started close to the claw hammer eye and walked towards the operator as blows continue. Repeat as needed to find the correct depth of cut.

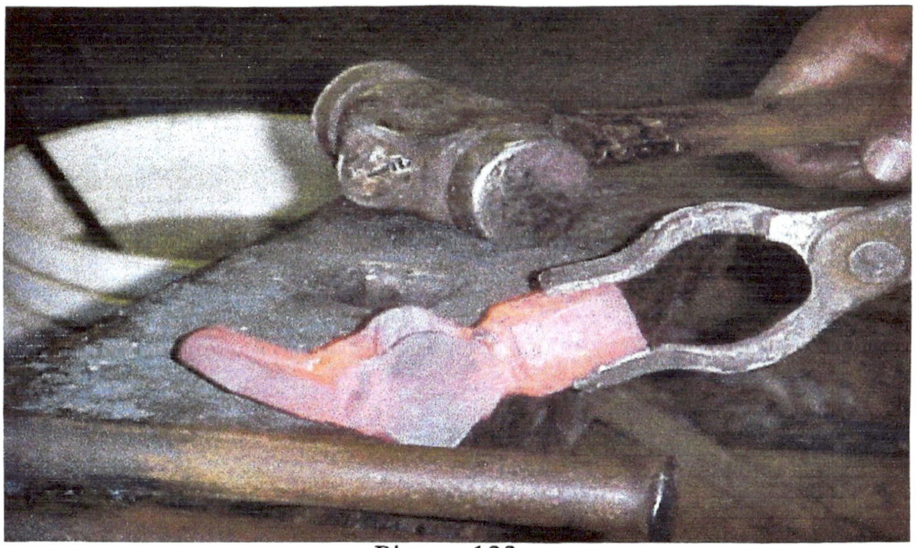

Picture 123

Our claw and eye are checked for alignment with the striking surface. From the ends, side, top and bottom views, the smith can train his eye to see flaws or factors that are going well. This claw hammer is now ready for a name stamp and type of steel engraving. We have been fond of signing each tool when it shows the level of quality sought after. The type of steel can be marked on a tool to ensure the appropriate heat treatment is implemented. We are also able to map which types of tool steel perform exceedingly well over the years. All engraving is done before the hourglass shape of the tool eye is completed.

Picture 124

Checking the straight drift selection to find the one which fits perfectly. Our drifts are not only used for eye elongation (as a mandrel) to prevent collapse, but also as a mandrel during engraving. A long drift also makes an excellent handle for tool manipulation. The taper of the long eye drift shown is 3/8 inch thick up to ¾ inch thick over a two foot length span. The last one foot, which is hand held, is completely round. Total length is three feet for this particular drift.

Picture 125

The hammer blank is forge heated to deep cherry and engraving can begin. A hardy hole fitted bottom tool is supporting the project to prevent tool distortion during this portion of our project. Engraving can be done on both sides of the forging by using this method.

Picture 126 – Finished Driving Hammer

As the forging winds out for the day, this is the hammer we were looking for; the last one. Note the clean vortex of flame from the fire. It should be as good at the end of the day as in the beginning. Often times a better flame is present in a gas furnace as the entire hearth and fire brick are now radiating heat to warm your billets uniformly. No cold spots present, so to speak!

Picture 127 (see opposite page)
Shown is a batch of farrier's driving claw hammers, they were all done on open flat dies without gauges or stops. Training our eyes to see errors before or as they are occurring takes some repetition to develop, but uniformity of results will then occur. All of the hammers pictured are fresh out of our wood ash can from their annealing stage.

Picture 128 (see opposite page)
Annealed cross pein and straight pein hammers all to finish at 2 pound weights. Also done with open flat trip hammer dies. Note the hourglass taper to the through holes.

Picture 129 (see opposite page)
Side view of forged farrier driving claw hammers, made of 1-1/4 inch hexagonal jack hammer bits. The larger ones will finish to a weight of 14 ounces. The smaller ones will finish to 12 ounces. The round steel chunk to the upper left is heated and placed first into the wood ash annealing can. Because of the small size of these tools, they will not hold enough heat on their own to fully complete a good annealing stage. This piece helps the first one or two tools into the large can maintain their heat until the others join them.

Picture 127

Picture 128

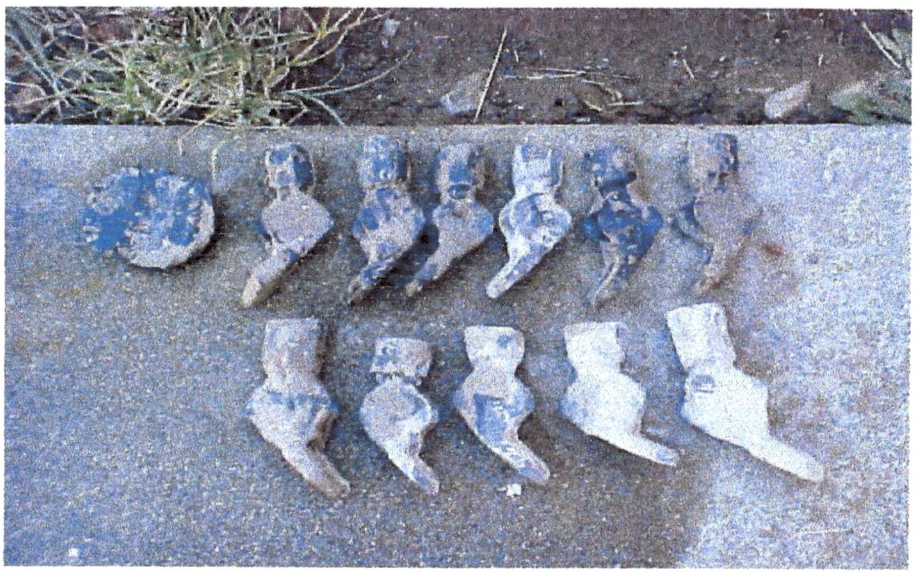

Picture 129

Picture 130
Side view of straight and offset pein forging hammers. All of these have a different degree of offset to the pein side which is achievable in open die or in hand forging (with or without a power machine).

Picture 131
We are looking at a modified radial arm saw used to cut claw hammer claws. The actual slit and/or V cut in this type of hammer is achieved beautifully and reliably with this set up.

A box leg vise is attached to the saw frame with two bolts. These bolts can be loosened and tightened to allow the claw hammer in the vise a straight cut shot from our saw blade. The cut is made into the vise jaws airspace, not across them.

Picture 132
Once the claw center is pointing straight at our saw blade we are ready to start. The blade used is a .035 speed cut abrasive wheel--very narrow and slightly flexible. An initial cut is made straight through the claw to the hilt of our underside fuller cut. The hammer can be taken out of the vise and checked for centering. The hammer is again secured in the vise and the saw started. Safety glasses are worn for all of this. The saw blade is again run to the end of our initial cut but as the blade is returning to us, the radial miter cut lever is pulled. This enables the entire wheel and saw motor to rotate freely. One side of the cut is now widened on the backstroke. Let the blade lock to center and slide into the initial cut end only once more. Pull the miter cut lever and begin retrieving (backstroke) the blade again, twisting the unit only slightly (approximately 1.5°) to dress of the opposite claw side of center line. Finish to the desired width. This abrasive wheel will cut annealed and hardened tool steel 1/4 inch thick with ease.

Picture 133
The base of our radial arm saw is shown. The bolts are to secure and adjust our box leg vise bracket to ensure alignment. The holes beneath this bracket are oblonged so that the vise can pivot forward and back; to and fro. Large washers beneath the mounting wood shown enables locking the bracket in any position which sights the vise jaw contents as ideal.

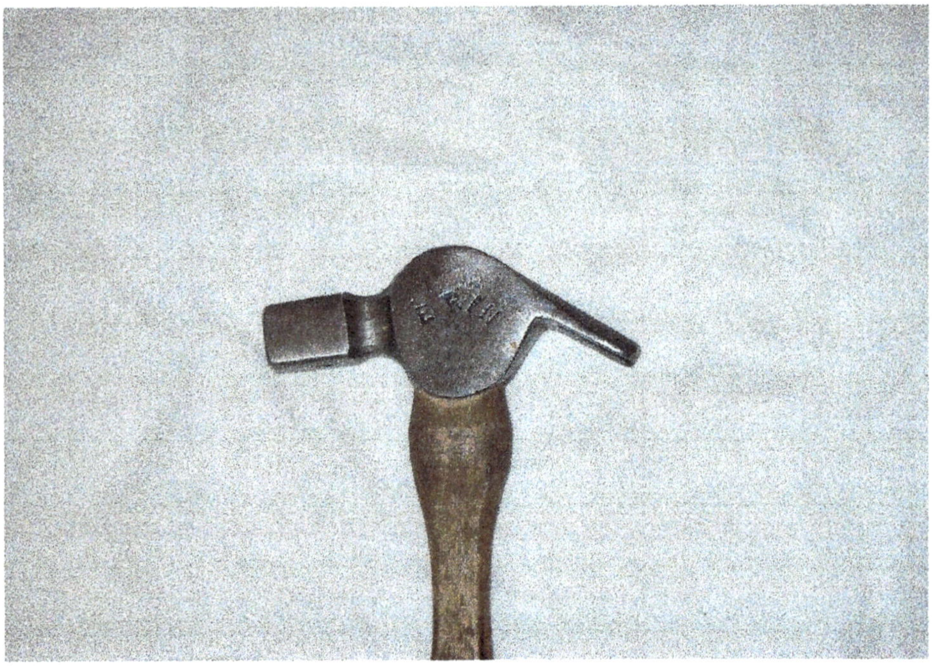

Picture 134

The finished driving hammer for horseshoeing. Ten ounce head weight and fourteen overall. Made from 1 ¼ x 4" hexagonal jack hammer bit remnant. Annealed, water hardened and tempered to straw color. A fourteen inch hickory handle is rasped down for comfort while leaving a prominent shoulder. A quarter inch deep by 5/8 inch wide hole is drilled into the handle end. A dime size magnet is countersunk and epoxied into handle end for nail retrieval. Note the angle of our face and claw; critical! The rule of thumb is to have the striking face flat on a level surface when your handle end touches. This applies specifically to horseshoeing applications in design. The process where form needs to follow function. For carpentry type claw hammers the standard has been to keep the striking surface plane closer to parallel with the handle shaft mid-line when viewed from the side.

There have been claw hammers forged here, which were tipped forward to have the striking surface approach parallel to our handle shaft. The head weight, distance from the shaft and dexterity of its user has much to do with the hammer face placement. Remember, this is a custom tool designed for a specific use and person. When used alternately with a generic type some true differences will be observed. The striking surface need only be forged or dressed 1-2 degrees up or down to create major changes!

The size of the magnet in the handle end also serves as a counterbalance. The type and weight chosen can alter the hammer's feel when swinging significantly.

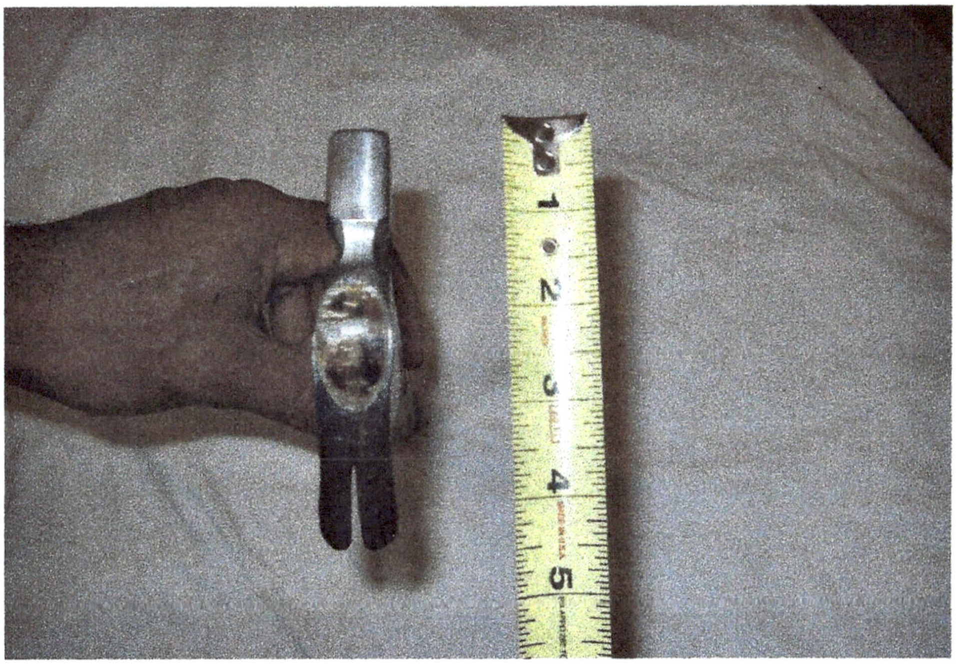

Picture 135

An overhead view of the same claw hammer. Note the clean spring fuller cuts behind the striking surface. The sharp claw slit with gradual taper for ringing off horseshoeing nails. This claw configuration also works very well for removing carpentry nails.

WOODEN HANDLE FINISHING: FITTING TO TOOL HEADS

Picture 136
Hickory handle is wrapped with heavy split chrome leather before securing in our box leg vise. Bracket in any position which sights the vise jaw contents as ideal.

Picture 137
The leather helps prevent scarring of the wood from vise jaw compression. Less pressure will be needed and greater security also results.

Picture 138
Upper wood to be inserted through hammer or set tool eye can now be rasped to fit. The handle is rotated and resecured in the vise so that all sides can be accessed. Rasp shown is a standard farrier hoof prep type. The coarse side is used first then turned over to fine side.

Picture 139
An old sander belt can be turned inside out here. It can be cut in half or kept entire. One-half is held in each hand. It is then pulled upon; one side at a time as a sea saw effect. This gives a great finish to a complex curved surface. Shown is 2"x 3' length of 60 grit shop roll.

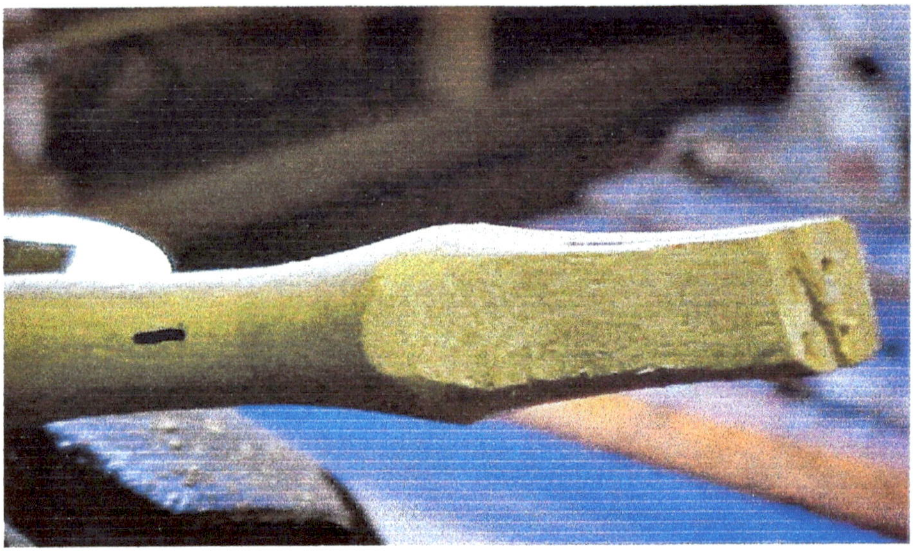

Picture 140
The handle has been reference marked to ensure same side insertion as hammer head fit rasping is continued. Most of our through handle wood is made square, then octagonal, and lastly oval or flat side oval. This helps ensure unequivocally that your hammer head remains centered on the handle. We are now creating balance for the swing of the tool.

Picture 141
Note some punch marks on the brace sides of the hammer head. These were made during forging and can be used to align the head on the handle.

Picture 142
Handle for this offset pein hammer has about reached its final destination for height through the hole. When driven through to seat, the right amount of wood for wedging will emerge.

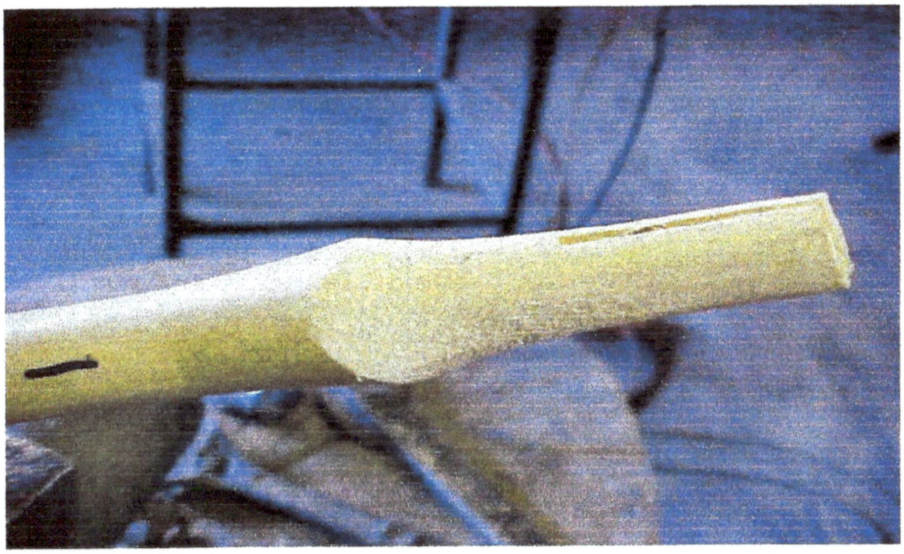

Picture 143
Flat side eye handle here has been fine rasped and is ready for shoulder sanding (see-saw method). It measures 3-1/4 inches from tip of shoulder to top where wedges are inserted. The longer the through hole, the more crucial is the time spent on inserted wood. The handle security will remain much longer over its useful tool life.

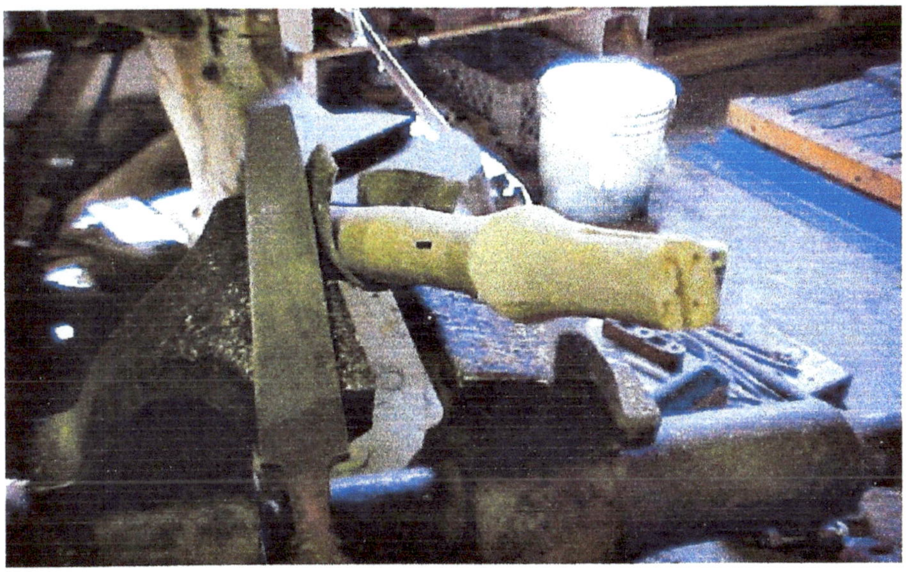

Picture 144
Prominent shoulder of wood handle is left on all sides below the swinging and set tools.

Picture 145
A 12 inch mill file with sharp edges can be used to smooth off exposed wood fibers after coarse rasping is completed.

Picture 146
Only the bottom portion of the through hole wood is finish sanded. Most all of our swinging tool handles are given an epoxy treatment after driving into the tool hole. Our epoxy adheres better to a coarse surface than a smooth one.

Picture 147
After all handle through wood is prepared, a 1-1/2 pound hammer is used to drive the handle in. It is done in mid-air as shown. Care should be made to keep the blows centered so as to not splinter the bottom edges.

Picture 148
The head is placed intermittently on the handle to check our progression to fit.

Picture 149
Note a hint of daylight showing on all sides of the handle as it seats.

Picture 150
Hammer slung from a box leg vise jaw with support leather beneath. The leather allows for extracting a handle which wasn't quite ready to drive through the head. Suspension of the head is only on its ends and swings freely (beneath faces).

Picture 151
The leather prevents scarring and scratching of the head as the handle is driven out. Note how through head hole wood has not quite emerged through.

Picture 152

The leather was cut (bottom left corner of the picture) from its perimeter to meet a 2 square inch hole in its center. This enables very quick handle insertion.

An old hammer handle that will fit loosely into the head eye is used to drive the new handle out. It only needs to be tapped by a light hammer from above.

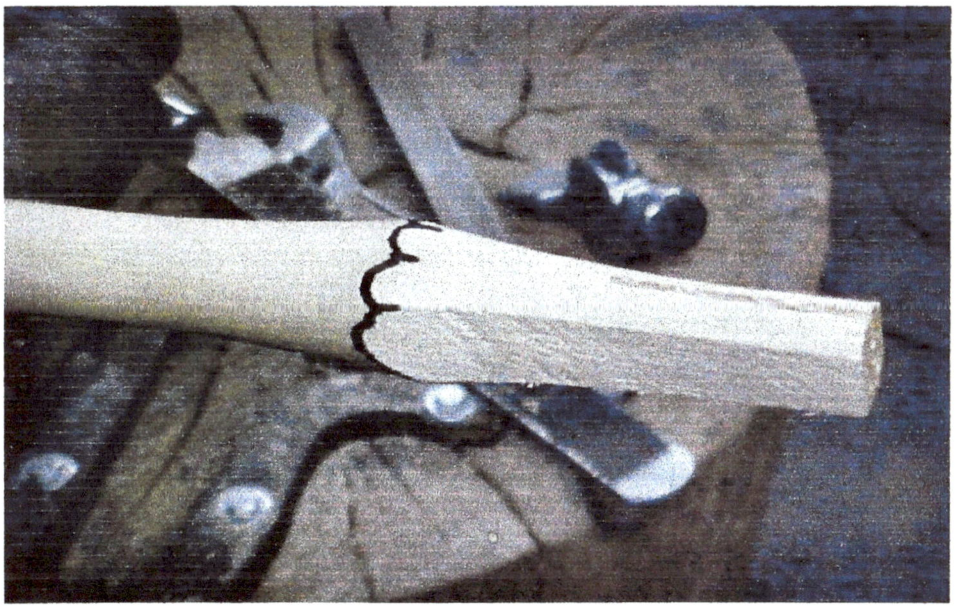

Picture 153

The handle shoulders can be dressed to include a flower petal design. This one was completed with a farrier's rasp. Held vertically on a round wheeled belt sander, against a tool rest, the same results can be duplicated.

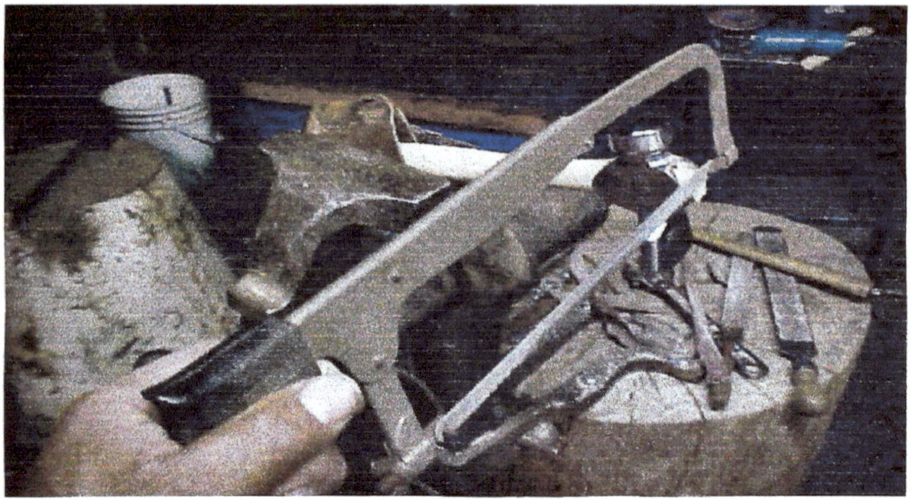

Picture 154
A box leg vise is holding a leather wrapped handle. The excess wood is cut from the upper handle. It is not necessary to have half an inch or more protruding beyond the tool head. Check that the wood wedge slit still has enough depth lengthwise down the handle.

Picture 156
This rounding/combination cross pein hammer has a good and true eye. The above (left) pictured hammer will need just slight trueing of the inner walls to maintain equal latitude of the handle. The longitude of the eyes of both hammers pictured (eye location, working face to working face) is perfect by our standards.

Picture 155
Position of a cross pein held in a box leg vise. The pein side is wrapped with leather to prevent scratches from occurring by the vise jaws. The eye can now be filed (chain saw file) to remove heat treatment slag and scale. Entrance and exit hole edges are also made blunt and rounded. A die grinder can also be used when the head is positioned this way.

Picture 157
A good gap between the through wood and the internal eye sides of the hammer. Equal distance on both sides of the handle is essential to keep the head orientation true after wedging. This amount of upper handle gap borders on excessive, but believe me when the wedges are driven to spread the handle wood only severe abuse will slide this hammer head off.

Picture 158
A nice view of the amount of wood we expect protruding from the hammer exit hole before wedging.

Picture 159
Handles having a damaged wood wedge slit can be repaired by rasping an additional "V" cut in them. This enables wedge entry without damaging either. Cutting the handle top shorter is not needed.

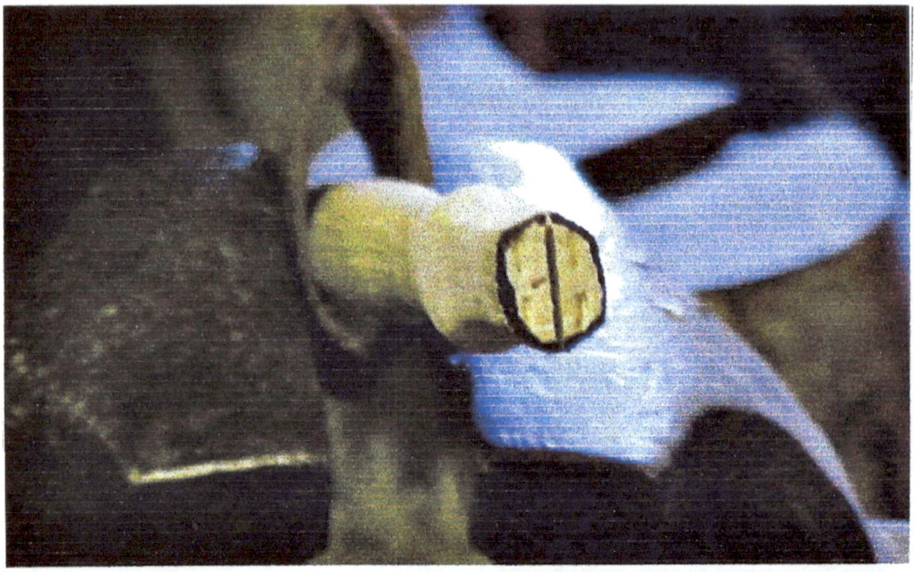

Picture 160
This upper handle edge has been traced with a magic marker. Each handmade tool eye will have its own unique circumference shape and diameter. The image traced and pictured here is an exact replica of the eye it will fill; just slightly smaller.

Picture 161
With the hammer secured vertically in our vise, the wood wedge is checked for width. The one shown is slightly oversized and will be rasped down to become narrower.

Picture 162
Epoxy glue is dripped into the voids between the handle and hammer head. More is placed on the wooden wedge. It is then hammered in until bottoming out in the handle slit. Excess wedge wood above the handle is nipped off before the glue dries. Good ventilation is employed and/or approved mask is worn when using volatile glues such as epoxies. Also, safety glasses are continually worn!

Picture 163

Excess glue that is squeezed out from wood wedge insertion is quickly wiped away with a clean paper towel. A small steel wedge is dipped lightly at the primary tip in epoxy and then hammered into the upper handle. Steel wedges are aligned against the linear wood grain. This helps inhibit a total linear handle split. An angled plier is used to hold the steel wedge as it is hammered all the way in.

Picture 164

An overly wide wooden wedge that is to be cut and rasped to enable insertion for a straight pein hammer. Occasionally a handle top may end up slightly smaller than is ideal (where it exits). Now is where the wood wedge fit can help correct this. Rasp the wedge down until it just fits into the hammer head eye rather than the length of your handle slit measurements.

Picture 165

Front view: Glue is applied first into the void surrounding the upper handle and hammer eye exit. It is not necessary to completely fill this void as the wedge displacement may make the glue overflow.

Picture 166

Rear View:
The same cross pein hammer at this stage. A small amount of glue can also be applied after the wood wedge is driven in to complete the seal. Wood and steel wedges have both been driven into the handle. They too were buttered with glue before insertion.

Picture 167

For this straight pein, the steel wedge was placed offset from the handle midline. It is set against the handle grain closer to the hammer's flat face. The one or two ounces moved to this side of midline helps balance the two striking surfaces. They are now made closer to equal weight on each side of the handle center (balancing).

Picture 168

For handles which are very smooth and have a fine tight grain structure, it may be difficult to see and remember where the lines are running. Wet the handle lightly with a damp towel dipped in water. As the lines appear, sketch their direction with a magic marker. These are the best handles to use for their tight grain is tougher. Note which way the steel wedge is inserted as shown.

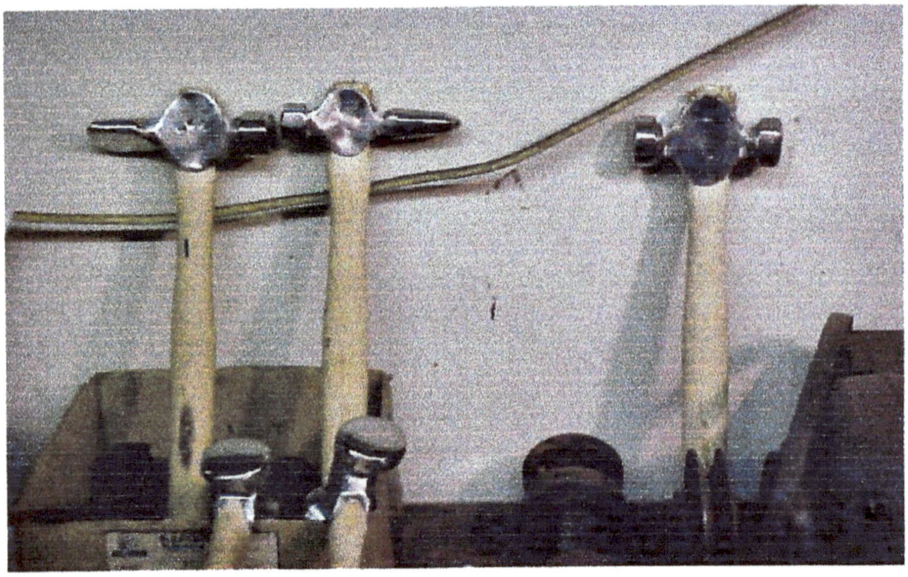

Picture 169

Hammer heads with all their steel and wooden wedges set with glue are placed aside to dry. Those in the near ground pictured required leaning. Exuding glue from compression will now flow to voids which would make best use of its presence. Most viscous glues are gravity specific and will flow to areas of least resistance before drying is complete.

Hand hammers of all types are leaned or positioned to dry. Their epoxy is applied and the drying position is dictated by the direction which is needed for flow to fill any voids. The epoxy used, as are many types of glue, is gravity specific until dry (flows with the direction of gravity).

Picture 170

The glue has filled all apparent internal voids left after wedges were driven in. It should not flow down any external surfaces as shown.

Picture 171
After allowing the glue to completely dry, the crown is dressed off. Here is a worn out 50 grit sanding belt saved for its last mission. The belt will usually load up with wood and glue, but this is its last chore anyway.

Picture 172
Many positions of placement for your tool to the belt will be needed to dress the entire crown. Protective eye and breathing apparatus should be worn as many small airborne particulates are usually dispersed.

FITTING TOOL HEADS: LARGE EXIT HOLE FLARES ALIGNMENT CONSIDERATIONS

Picture 173
An offset cross pein hammer rested on the edge of a wooden stump. The handle end above is lightly tapped to achieve a parallel aspect for the steel working surfaces and the wood handle mid-line. When handles are installed by hand, this is done before the handle is driven to its final destination.

Picture 174
A good look at an offset cross pein with exit wood from the handle being the focus. The amount and height of wood showing above this hammer is about ideal and ready for wedges. Just the right set up for allowing wood and steel wedges to spread our handle. The split shaft end will fill the hammer eye hour glass shape.

Picture 175
The offset cross pein mounted in a large box leg and ready for its final security wedging. The handle shaft has been wrapped with split chrome leather to prevent scarring from the vise jaws.
Resting the handle end on the vise through collar is the best placement. Bench mounted machinist vises will also work for this installation. Just let the handle end find your bench top before cranking the vise screw tight.

Picture 176
Close up of split chrome leather wrap on hickory wood handle. Handle end is resting on the vise screw collar. Lastly, our vise screw is tightened.

Picture 177
The upper edges inside our wood wedge slit are being rasped at an angle. Occasionally the slit is compressed closed to some degree when driven through a tool eye. Whether a set tool or a swinging type, this slit closure is circumvented by rasping a "V" cut into the cleft remaining. A wood wedge can now find its way into position without breaking.

Picture 178
Single wood wedge installation for a swinging hammer. This slit had closed completely and was rasped to a "V" cut before the wedge could even begin to enter.

Picture 179
A double wooden wedge set up. At the time an abundance of wood wedges not quite thick enough to spread this handle's wedging slit. Two smaller (thinner) wedges of the correct width and length were epoxied together. Their outer surfaces are also coated with glue before being driven into position.

Picture 180
A lateral view of a double wooden wedge set up. The width of these wedges have been rasped to allow for entry to the fullest of this hammer eye.

Picture 181
Double wooden wedge set up materials on a disposable cardboard remnant.

Picture 182
The epoxy glue is mixed on our cardboard and applied to one side of each wooden wedge. The wedges are then pressed together with glued sides touching. Good ventilation and/or a breathing mask should be ever present when using any volatile glue.

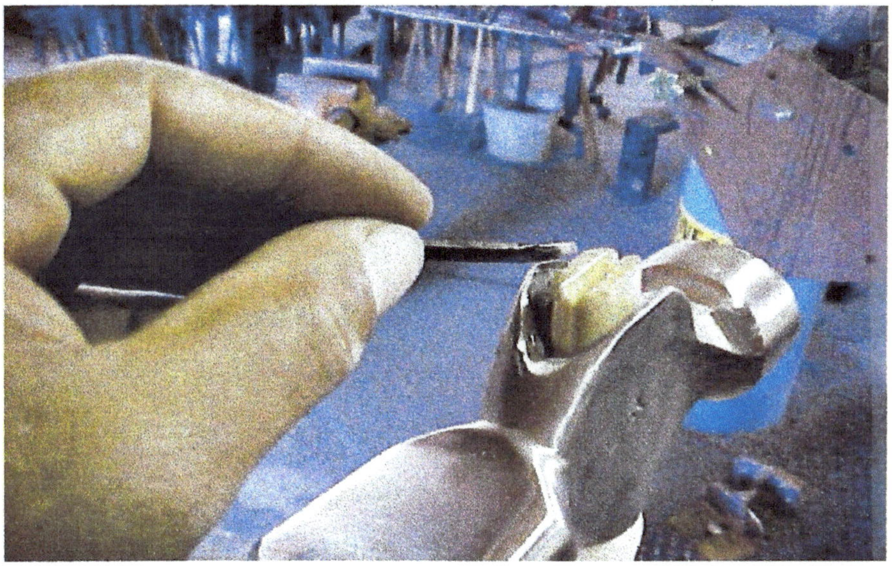

Picture 183
Before our glue begins to show any signs of solidification, it is dripped into the voids surrounding our exit wood and hammer eye.
These voids do not need to be filled up to the brim; for as our wedges are driven inward, an overflow will result. After some practice, the smith will become familiar with an appropriate volume of glue and ratio of eye void: wedge size displacement.

Picture 184
Our wedge sides are lightly coated with epoxy. It is now driven inward and excess glue wiped away with a clean paper towel.

Picture 185
A piece of mild steel bar stock long enough to be hand held and wide enough to cover our complete width of the wooden wedges is held above the present insertion. Another light (1 lb) hammer is used to tap the wooden wedges to bottom out. Wedges which protrude well beyond the tool eye even after reaching the slit bottom will not need this step. The wooden wedge is merely nipped off (not sawn), as the glue is still viscous and will make a mess.

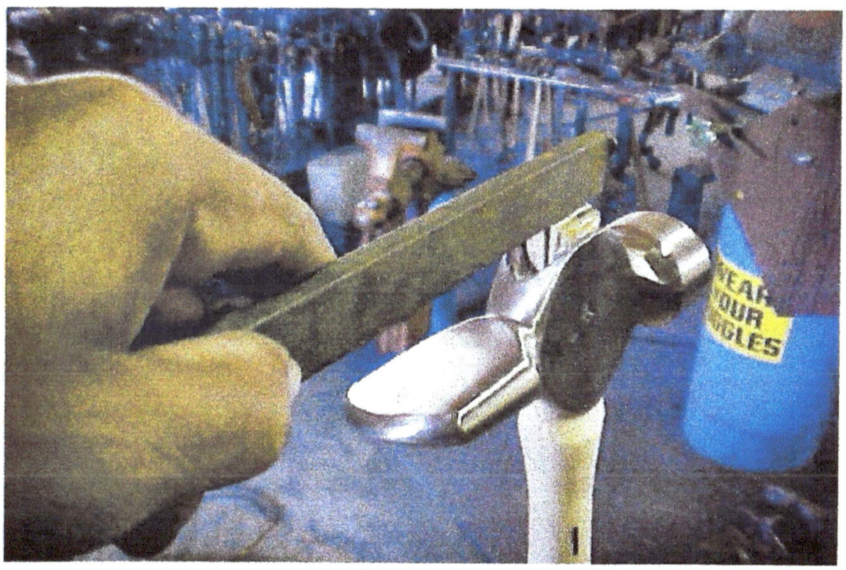

Picture 186 (see opposite page)
After the wooden wedge is set and before the glue solidifies, a steel wedge is made ready to install. First the wood grain of our handle is located. Its longitudinal direction should be located as the steel wedge will *need* insertion opposite to the handle grain. The wedge size (length and width) should be proportionate to the mass of your tool head, size of the eye, and the type of wedge. There are many types of steel wedges: tapered and serrated, corrugated, oval tapered tube with or without serration, smooth steel or aluminum, crescent moon, etc.
This wedge is held with an angled jaw pliers before hammering into the eye wood. A needle nose tong will also work well for this operation. This wedge is not held by our fingers for it too is coated at the primary tip with epoxy glue and our digits should be kept clear to drive this wedge.

Picture 187
Once the steel wedge has been driven in as far as is reachable with a hand hammer, additional depth is made possible through the use of punches; most often a mere tap or two against this punch as it rests on the steel wedge. The point being just to set this wedge at or slightly below the eye exit hole level when lateral viewing our tool. Leveling the wedge is easily achieved with this tool and many other set tool punch configurations. All voids should now be full from wedges, glue or handle wood.

Picture 188
Set punches used to finish setting our steel wedges have narrow and blunt (flat tipped) working ends to correspond with each wedge thickness.

Picture 186

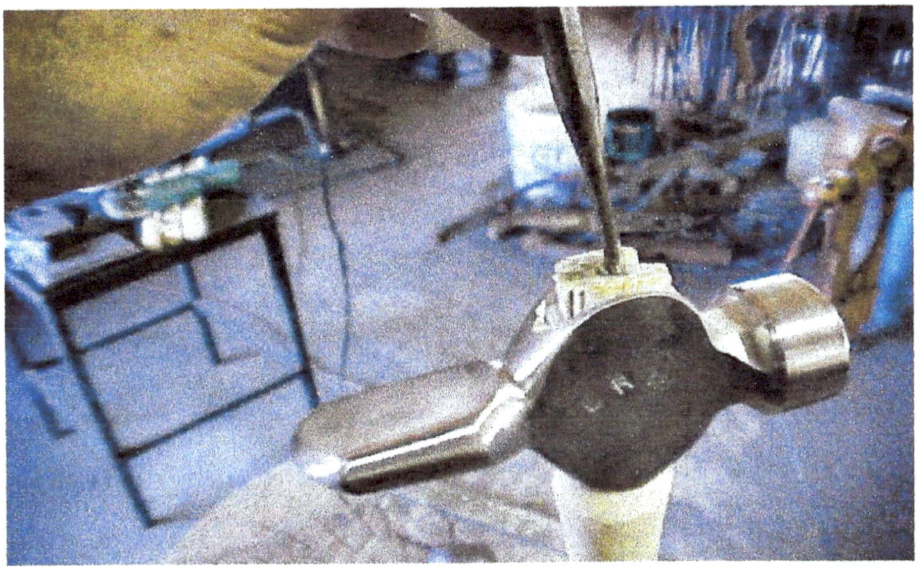

Picture 187

Picture 188

Picture 189
Occasionally a tool handle is set as desired, yet excess glue continues to exude from the eye due to the forces of compression. The wet glue only needs something to help it dry in place. The tool head is stuffed - exit eye down into a bucket of hickory sawdust. It is then placed upright as shown to continue the glue drying process.

Picture 190
Small remnants of unused wooden wedges can be spliced together too for one useful unit.

Picture 191
The wedge pieces all have the same length and degree of taper. Their widths have been rasped down to fit this hammer eye when assembled as shown. Saving unused pieces can be useful.

TOOL FINISH WORK

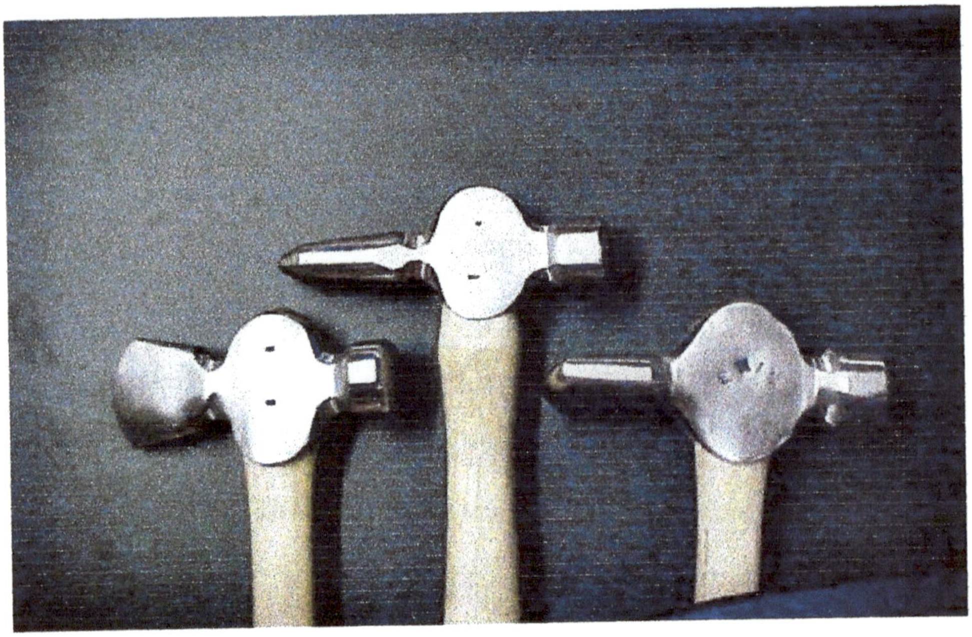

Picture 192
Left to right, freshly crown dressed straight pein, ball pein, cross pein ready for use.

Picture 193
A close up picture of the 4-1/2 inch flapper disc of 50 grit. Keeping one of these around the shop even for occasional use can be indispensable. Many industrial supply houses will carry innumerable sizes (thickness and diameters), hardnesses (flexible/rigid), and grits.

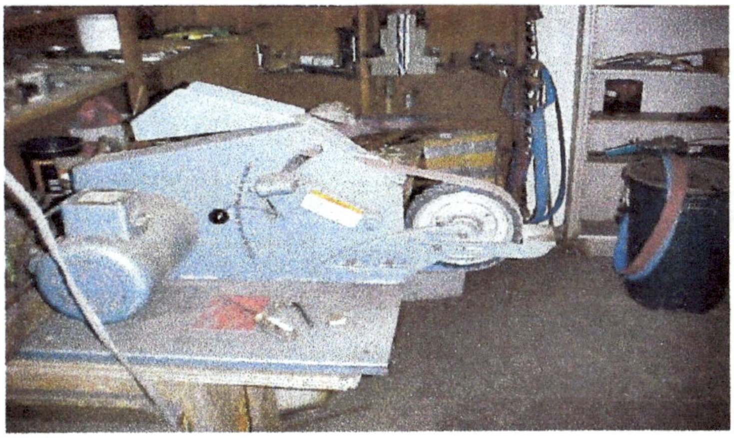

Picture 194

A good example of one of our finishing machines. This one is known as a square wheel grinder. The fittings and accessories which can be changed and incorporated enable inside and outside radius grinding. The grinder belts can be changed to many different grit (abrasive) combinations. The tool rest can also be moved to accommodate safe usage. Safety eye glasses, heavy gloves, and breathing mask are worn when in operation.

The pictures shown are all of the same machine. The fittings have been changed out to show some of our options possible.

Machinery of any kind that will perform reliability is not inexpensive. Why not get a good one which can serve several functions over many years?

Pictures 195 & 196

Design, fabrication and construction of good finishing grinders is possible. The Smith must weigh many advantages and disadvantages of time devoted to this project. Will the time of construction required equate to a similar or better finished product? What factor is in least supply - time or funding? Are there any time constraints? Example: a tool which needs finishing in a fairly short time period. Can this unit serve for many future challenges and present cost is not the greatest factor? Does the type unit needed even exist in the desired form, or will one have to be fabricated? Can some portions be purchased and others added or modified as requirements dictate? Does the smith even need such larger machinery or admonitions for daily operations;

Would they do better to farm out that segment of work to another tradesman?

Picture 197
Trueing up any irregularities in circumference can be made both before and after heat treatment. The heaviest of which is done prior, as is head balancing.

Picture 198
Areas exhibiting a miss hit from the forging operation are dressed off for better eye appeal. Often when blemishes are away from the striking surfaces, many smiths may say "no harm, no foul". Yet, why not simply get rid of their trail?

Picture 199
Dressing away eye exit hole blemishes and nicks will make for a safer tool when in use. A swinging and set tool should have no edges sharp that would cut or scratch the operator.
Safety first and foremost...

Picture 200

Occasionally a cross pein hammer requires trueing up to adjust the pein and handle eye misalignment. When at a deep cherry heat, this is how its done after forging is completed. A long straight drift matching the through eye dimensions is inserted into a vise secured head as shown. A push or pull on the long drift is all that is needed. Cross peins are not made into straight peins by this method as a severe shear can result between pein and eye shoulder union. This is done only to correct any small 1-2 degree imperfections and keep a tool from finding the "uh oh" box.

Please take a moment to ponder this … the pein holds still, the eye and broad face working surface moves. In less common instances, I will hold the eye (with the drift in place as shown) and move the pein. For this type of forging correction/trueing, the chief element to remember here is that one portion must hold still, another can then be moved.

Picture 201 (see opposite page)

Shown is fresh batch of various peining hammers. Straight, cross and offsets are laid out for final grinding and polishing after heat treatment is completed.

Picture 202 (see opposite page)

The pein striking surfaces can have many different constituents incorporated to suit the smith's personal preferences. They include weight of pein, length, width, degree of offset, direction of offset, left or right handed use, pein crown radius. Each aspect is now finely tuned to meet requirements for that tools intended use.

Picture 203 (see opposite page)

A muslin wheel can be loaded with many different types of rouge. For our purposes, the rouge is fed into the wheel and allowed to dry. The product selected should be of an appropriate grit.

To aid in our finishing process, the grits can range from 80 to 2000 with all points between. Some formulas of rouge will produce sparks from the steel being finished when in use, others will not. There is no better or worse here, simply pick one which allows efficient smoothing and blending of your work without excessive bearing force into the muslin wheel. Over time and use, the old rouge remaining on this wheel may become pitted and cracked, making it difficult to reload additional abrasive to the wheel. Our preference is to scrape old crusty, unused material from this wheel using an old worn out file. While the bench grinder is turning, put on your safety glasses and leather gauntleted gloves. Present the file working tip to the worn out rouge to peel away old material. This shouldn't take much time or pressure. As the old rouge is dislodged, fresh muslin will be exposed. Turn the grinder switch off and present the next formula of rouge to the muslin wheel as its revolutions wind out. When the wheel stops, check to see if sufficient rouge was picked up to be put back in service. Repeat this last step if needed without the file scraping.

Picture 201

Picture 202

Picture 203

Picture 204
A close up of our muslin wheel freshly loaded with Lea C Rouge 80 grit compound. On the side of our wheels, an arrow is drawn to show the direction of wheel operation. When a wheel is removed from a machine there will be no confusion as to which way it was revolving. A big plus for our polishing operations and other grinder wheels as well.

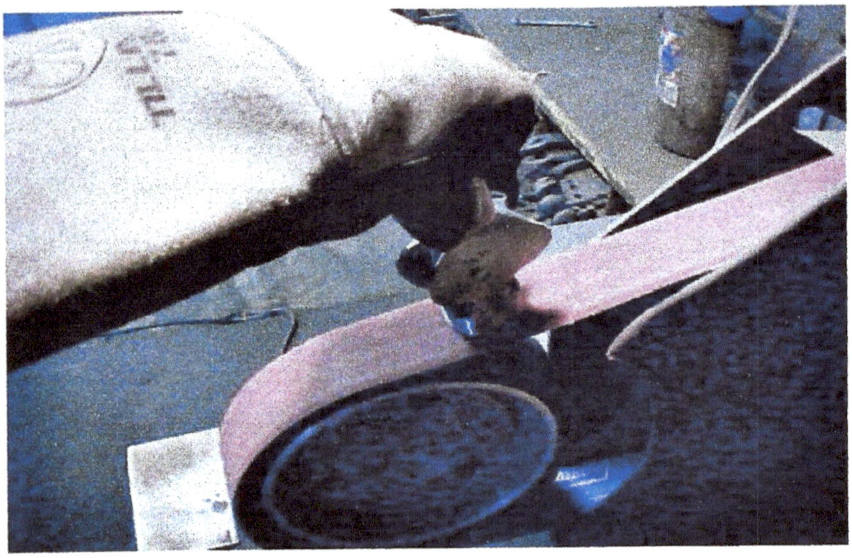

Picture 205
Finding a means of dressing a hammer surface flat or slightly crowned, can be a healthy chore. Holding a head lightly while allowing gravity and a slight friction from an abrasive belt to show our eyes and hands where the high spots are. Rotating each head only a few degrees at a time takes out much of the guess work. The head is but floating on our abrasive belt; no pressure applied to find these.

Picture 206
Hammer head striking edges are given a final contouring before their last polishing. A 60 grit belt shown in use.

Pictures 207, 208, 209, 210, 211

How a tool head can be checked for good balance. For many of our swinging tools, ideal balance is considered equal weight on either side of the handle. Some smiths may differ from this for specialty tools. Even most industry standards such as ball peins prefer the pein side to be lighter. Here the smith must decide what feels right and helps deliver consistent and accurate blows. Many set tools are more comfortably and accurately used when a greater proportion of weight is borne below the handle (working surface) and less above the handle (striking surface). Playing with this balance will help you become a better critic.

Shown are ways in which the smith can lay out a forging to be judged for balance.

Picture 207

Picture 208

Picture 209

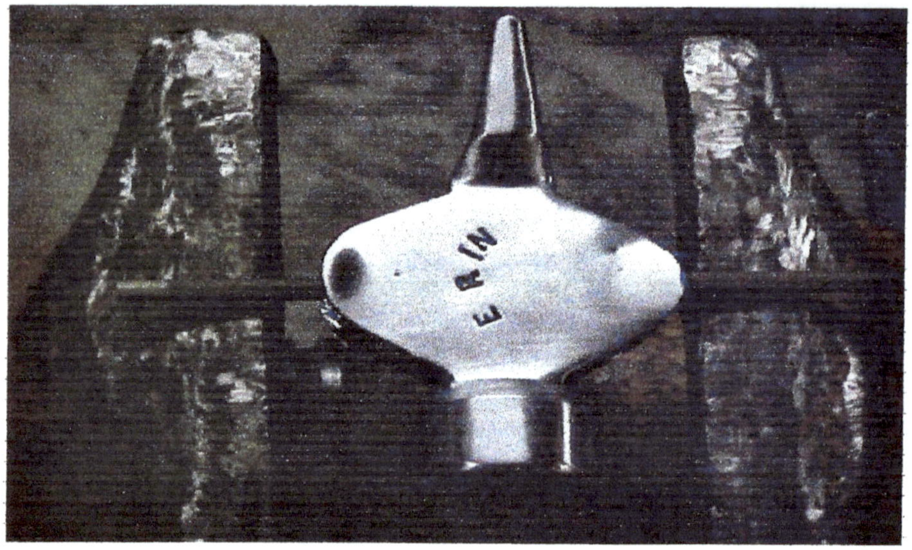

Picture 210

Picture 211
We call this one the
"Teeter Totter."

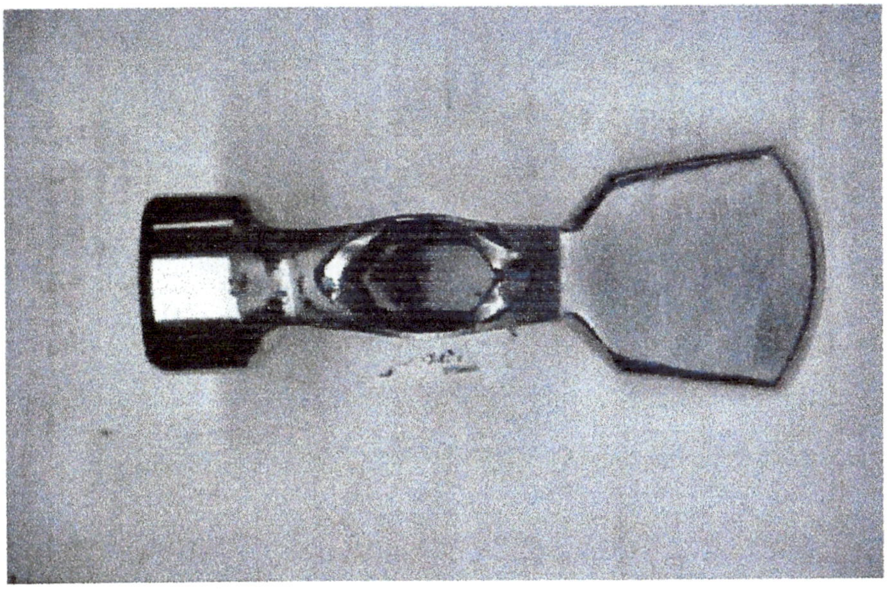

Picture 212
When the striking surfaces of a hammer are considerably wider than your eye width, there is a much greater chance of having uncontrollable blows produced. Each swinging tool will have what we term a "sweet spot". An area and corresponding surface area of so many square inches which will achieve great results when in use. Outside of these parameters glancing blows result. Handles and tool heads twist in our grasp after impact. Viewing our tools in this light will help find the limits of your tools "sweet spots".

Sometimes the problem is weight; which must be removed. Other areas are just volume- too much surface out where it cannot be used efficiently and gets in the way.

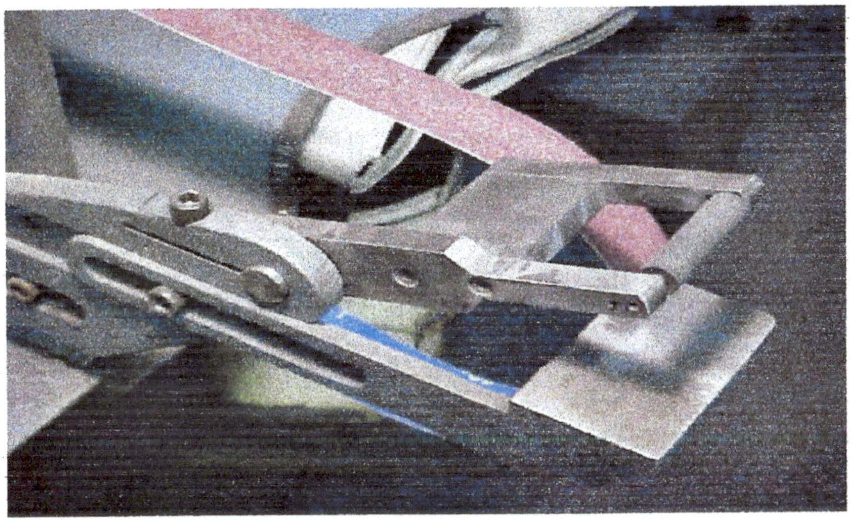

Pictures 213, 214
A half inch radius grinder wheel attachment used here to clean out fuller cuts for tool heads. Many different fittings for finish work are available, but the smith must seek out options.

Picture 215
Position for cleaning out fuller cuts, cross pein hammer shown.

Picture 216
Our 5/8 inch top and bottom fuller cuts were made to forge the hammer. The hammer billets were rotated to create an octagonal concave isolation.

Picture 217
The entire concave circumvention is cleaned out with a 1/2 inch radius grinder wheel. This is very narrow for a grinder belt so care in pressure applied is essential. A new 50 grit is used here to remove slag and residual heat treatment scale.

Picture 218
During this grinding, the process is ceased before oxidation colors of heat develop on the steel surface.

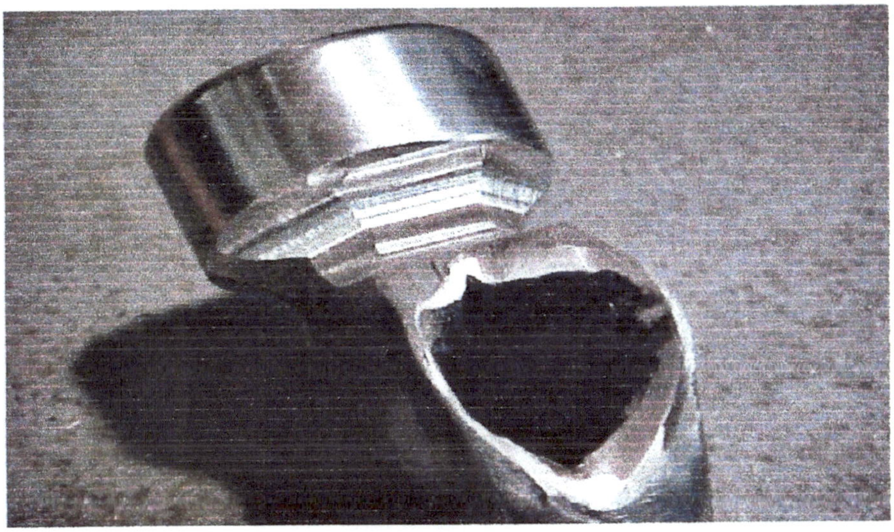

Picture 219
Fuller cut grinding is examined to ensure uniformity in depth and hence width.

Picture 220
Hammer heads are cooling in front of a large fan after fuller cut clean out grinding. This helps prevent heat of grinding from altering any of our hardening and tempering process previously completed.

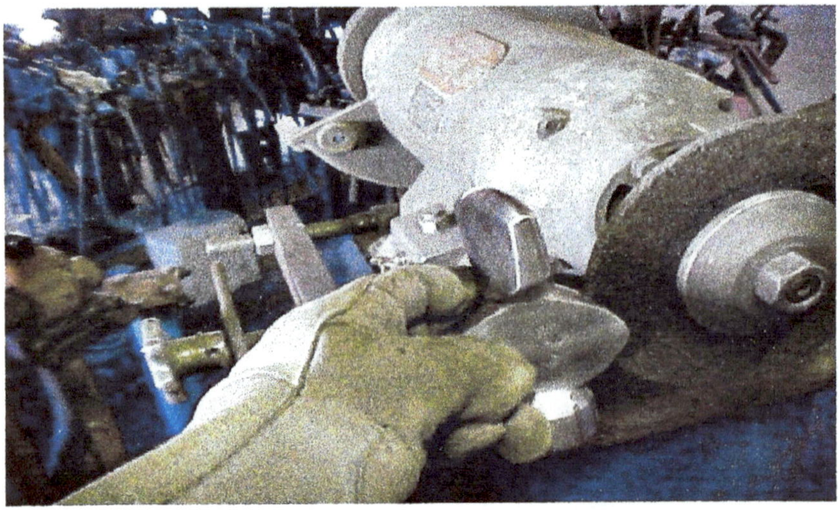

Picture 221
Here a 9 inch angle grinder 36 grit wheel is installed in a bench grinder. The wheel was purchased with a center hole to fit our bench grinder spindle. Using this configuration has helped in cleaning out internal edges of tool eyes. Note the heavy protective gloves being worn for this.

Picture 222
An end cap to contain all abrasive wheels of bench grinders is a real good idea, without which many lock nuts can become compromised in use or crush the actual grinder wheel during tightening. Even a large flat washer used as an end cap placed on before the lock nut is better than nothing.

Picture 223 (see opposite page)

An exposed view of a grinder spindle. Numerous flat washers have been placed on first to achieve distance between the grinder wheel and motor. A section of pipe which closely fits the spindle is next put into position over any exposed threads. This pipe acts as a collar to keep whatever type of abrasive wheel centered around the spindle. The wheel will now spin true and not eccentrically (off center). This is done for wheels that are too large in center hole for a grinder spindle. The pipe length should match only the abrasive wheel thickness. The pipe can be steel or plastic, which work equally well. Light weight, aluminum or brass, can be cut into small strips and bent around the spindle to form collars of many different thicknesses.

Picture 224 (see opposite page)

The center hole of a typical abrasive grinder wheel without threads. It is easier to fit this design to a bench grinder than some threaded varieties. The thread direction, diameter and number of threads per inch is not a factor because of this choice; not to mention the actual spindle and disc hole diameters.

Picture 225 (see opposite page)

A 9 inch flapper disc placed on a collar on our spindle. The next step is to put the end cap and lock nut behind it all. Flapper discs are great for many types of finish work, though, in a bench grinder care must be taken to be sure the rotational direction is correct. This grinder, as are many, rotate toward the on/off switch throwing sparks towards the foreground when in use. Be sure that the small pieces of unidirectional sandpaper on the disc will do just that without catching an edge of paper. These discs are undirectional (work only one way). They can be purchased to operate (rotate) opposite; to the far end of this type of grinder if needed. Grit types used are coarse to start (36-60) and finished with finer (80-120).

Picture 223

Picture 224

Picture 225

Picture 226
The bench grinder with an appropriate cover plate on the flapper disc. Now ready for its lock nut.

Picture 227
In this spindle set up, a flat washer which can fit over the collar it was placed on. This will act as a better backing plate than the smaller washers behind it.

Picture 228
This view shows how the larger washer provides a little extra support for the flapper disc. The cover plate (out of view) now has a surface to press against for security, without flexing the disc when our lock nut is tightened. The inside (backing) washers and outside (cover) plate can now oppose each other.

Picture 229
Here a fuller cut radius is easily cleaned out and made ready for its final polishing. The flapper disc can dress many hard to reach areas. It's one of those wheels which teaches the smith to make decisions. Should we hold the tool in our hands and present it to the grinder, or should we secure the tool in a vise and present the grinder to it?

Picture 230
An outside contour is being dressed on this cross pein hammer with the flapper disc in a bench grinder. Remember to hold onto our tools with both hands as much as possible. These discs are designed to operate by running in one direction only. Be sure to install them on the correct end of the grinder or with the correct rotation for use!

Picture 231
The flapper disc in a bench grinder can reach an internal (concave) radius where many conventional wheels may have difficulty.

Picture 232
A flapper disc in operation to smooth and contour the hammer eye exit and entry edges.

Pictures 233, 234, 235, 236, 237, 238, 239

The pictures shown are for the smith who has a desire to incorporate accurate finish grinding to their work. The implement turns towards that of a machine shop, yet all pieces are from the smithy scrap pile. This particular set up has no name and in no other text has one been found. Its function is similar to that of a reverse lathe for as the grinder revolves towards an operator, the work project revolves away (opposite).

The lathe's purpose is to clean out fullering cuts of hammer heads. A sliding tray enables complete and accurate grinding of the hammers complete circumference to a uniform depth. The chief requirement for this implement is only that the tools under construction have a through eye for a wooden handle, as most of our blacksmith tools do!

Picture 233

Picture 234

Picture 235

Picture 236

Picture 237

Picture 238

Picture 239

SHOP ORGANIZATION

Picture 240

Shop arrangement seems to be one of the chief stumbling blocks regardless of size. There must be a balance between workspace, that area where forging is actually done, and tool storage space, that area where tools of all kinds remain at rest.

It has ever been a challenge to venture to neighboring smiths to complete a simple task. An infinite selection of tongs, set tools galore but no possible way to walk from the forge to an anvil because of clutter. Some of the best operating shops are shorter on tools but great on workspace. In time the tool selection will improve if new ones are added from time to time. Over the years, each tool should find a rack or shelf or table slot to call its home. When a project is finished, the tool goes back to its home. Similar tools, if not for shape or size, but function in a forging process are kept near each other.

Picture 240

The aluminum silver box beneath the forge is a slide out type on rollers. Here are kept all the fluxes for welding and brazing steel, aluminum, brass and copper on a weekly basis. All those not used weekly or just in need of storage are kept on a dry shelf elsewhere.

A tool table supports one of our large shop forges. A tennis ball is sitting at its upper right side for size reference. The tong hanging from the tables lower left are 3 feet long and used in our small anvil making. The tong pairs in the lower foreground are all flat jaws. They are arranged from smallest to largest in jaw bite, left to right.

Tong rein lock collars of all different sizes are resting at the near right, a handy spot to grab and slide on a heavy tong before grasping a billet needing additional forging. Hardy hole shank mounted collars for rivet setting and loosening are resting at the near right.

Picture 241

At the top of this forge are burners and venturis which can be unbolted and changed out in only minutes. The burner pipes can be adjusted in terms of height to enable our flame vortex within the fire box to become centered on the billets. Our hottest flame portion will reach the steel. Over time, these burner pipes may become blistered (2-3 years) with daily use at a near welding heat. They are easily removed and replaced with this set up. They are made up of thick walled schedule 40 black pipe. The diameter of this pipe will need to accommodate your forge height and internal volume.

Each pipe diameter will have a different flame vortex dispersion in a gas furnace. Factors determining vortex characteristics (length and width) are flame exit height, length of burner, pipe diameter, fuel pressure level, and air draft mixture. The baffles (air mixture plates) to the rear of the venturis shown have the ability to control how much air enters our flame. These operate much the same as a modern day oxygen/acetylene mixing handle chamber works. The exception is that our entering air is not pressurized. As the baffle is choked down (closed), the flame desires more air to maintain its hottest vortex. The draft ensued will hence pick up speed. An adjustable draft baffle can enable many ranges of gas flame as well-blacking flames, blue, orange, yellow. The baffle is threaded so that minute increments can be slowly sought out. A lock ring secures the baffle when the desired setting has been reached.

Our own ignition system here is a spark plug into an automotive non fouler welded to the burner pipe. The wire running up to the spark plug comes from a piezo – electric button (similar to gas barbecues) which is bracketed below and well out of danger of being melted when the forge is opened.

The heavy steel loop welded between the burners is for lifting our unit by crane or tow truck. This forge with heavy liner detached from the table below weighs nearly 400 lbs.

Picture 242

Shown at center are our ignition button and gas regulation valves. The valves are of high quality brass fittings as are used on good gas welding sets. They can turn gas flow to burners up or down, on or off, separately or together, as required. The hoses and end fitting are for high pressure operation. These should be made up only by a trained professional or a shop which can properly affix the correct fittings. Pre-made hoses can be purchased and used only if their quality is good and their length will allow for plenty of clearance between the forge. As the hoses traverse from valve to venturi, the air space between hose and forge casing must allow for safety clearance. This is one area which should not be taken lightly, as a hose which becomes over heated (hose too close to flame or forge casing) may result in disaster. This can be true for shop built models or professionally available factory units.

The median (center) hose pictured is the incoming fuel line from our gas storage tank. This tank has a pressure regulator valve and safety flow limiter. Should a fuel line ever become severed, the flow limiter shuts the system down. A good feature to have on any gas fired system.

On the forge casing, angle iron is welding horizontally. This angle is large enough to accept ½ inch round stock. The round stock is then bent and shaped to fit forward and aft of the forge casing - right and left of the casing. The ½ round stock now makes a sliding extension in front and to the side of this forge. This enables long pieces to be heated without sagging of its colder portion protruding exterior to the forge. This type of extension works well on coke and coal forges as well, for they slide back and out of the way of work when not in use.

Picture 243

With a large forge in operation such as the one shown, radiant heat is dispelled when the front door is opened. A smith may tend to crowd the anvil or box leg vise nearer to the forge for expediting work, even to take less steps from fire to work station. But to open the forge, everyone must clear out to prevent radiant heat burning and temporarily stop operations until the door is again closed. Three or four walking steps is considered about the right distance from fire to forging station. For our operation, a one or two walking step (fire to anvil) is handy when using the side port of this of this forge, but too close when using its front door.

Working close to any large well insulated and contained forge may seem comfortable and handy as less walking is required. For a gas fired furnace that is opened, a wall of heat will exude into your workspace. Now the billet is hot and the workspace is too hot to work in. Shop arrangement should allow for this as each heat will provide the same result,

The forge station shown also allows for an overhead crane to wield billets and large tong. The billets can travel from fire to anvil or trip hammer as unobstructed as a straight line will allow.

A large free standing and oscillating fan is pictured to the rear. This fan is aimed to keep the workspace as comfortable as possible in hot or colder weather temperature days. Even in a shop which is completely self contained, the exterior weather will effect what can be done on a daily basis within a shop without prior proper preparation of layout.

Picture 244

When running off many tong pairs that will not be finish ground, riveted and sized immediately, they should be tied together with their mate. Whether your tong is of unmistakable design (could have only one mate), or is near identical to many previously shelf stored, they will be more easily retrieved and finished when tied together. Many a tong may need finishing to complete a separate project. Having your blanks stored this way will help expedite other aspects of the shop operation.

Picture 245

Two work stations within the shop. To the left, a deburring wheel in a bench grinder and low box leg vise in the foreground. The leg has been shortened on this heavy (200 lbs) vise, used often for upsetting, heavy bending, rough grinding and hot rasping. The height of the vise jaws has been made to equal that of the shops anvil faces.

To the right rear of the picture is a large oak stump used as an anchor, work table and ballast for a larger (250 lb) box leg vise. The operation at hand has been preparing hand hammer handles (wooden stems of osage orange).

The large vise to the rear holds the stems at a height *most* suitable for debarking and hand rasping. Note: All sawdust and electrical cords are all cleaned up and relocated before any forging is commenced.

Picture 246

Tool caddies and carousels can be full without causing clutter. All hand tools are then readily accessible when placed in an orderly fashion. Hammers of similar shape and purpose can be arranged small to large progressively. Top swedges can be arranged by size or function, small to large consecutively. Tools which are only used in conjunction with one or two others can be kept side by side. Example: brass hammer next to chisel hardies.

Picture 247

A good swedge block station. The 150 pound block can be rotated or turned vertically to provide access to all sides for heavy use. Swedge blocks are used for countless types of steel forgings -bending, shaping, smoothing, contouring, upsetting shouldering into the center holes, drawing, forming, texturing, etc. Not to mention swedging square, round, flat steel and many others. All completed by hand or with a striker (helper with a sledge hammer). The tool rest for this swedging station is attached. Some may elect to have it separate for dampening any noise during heavy forging. The blocks working height should be at or slightly lower than an anvil face height which is comfortable for a given smith. The stand should allow for this height maintenance even when the block is rotated or standing up vertically.

Picture 248

The bar stock shear shown here is floor mounted in its left side. The right side pictured has a 5 foot handle attached. The entire tool is made of mild steel bar stock 3/8 inch thick. The slots will shear 1/4 x 1, 5/16 x 1, 3/8 x 1 dimensions and smaller. The slots were cut with a simple chop saw and abrasive wheel to the sizes and depths described. Mounting holes were drilled through the base plate half. A larger drill bit was used to ream the two mounting holes to one half of the plates thickness. This is to allow the mounting bolts to become countersunk when tightened. We now have a completely smooth bearing surface against which to shear steel, brass, copper, aluminum, etc. This shears action is via twisting. The base holds still and the handle moves.

Picture 249

A rolling tool rack designated to hold some of our hand-held power hammer implements. Resting close to a machine that is used daily, items can be seen easily and the desired tool retrieved, all within a half walking steps reach. The working ends of each tool is pointing up. Tools which were just used and need time to cool or be quenched are laid elsewhere.

Picture 250

A 2 inch OD (outside diameter) schedule 40 pipe steel rack. Organizing reserve materials is essential to helping keep shop operations neat, tidy and safe. Elbows, feet, knees and hips all seem to find misplaced materials not stored away from the workspace. When a project is in midstride or near finished, putting away unused items or remnants can save many accidents.

This rack was constructed completely of old pipe with coped ends before arc welding together. A little spray paint in a ventilated area and it cleaned up beautifully. It was made light enough so that one person can move it around empty without help. This is a free standing unit which takes up very little space as compared to the material it can hold.

THE PUNCH EXTRACTOR

Picture 251
An eye punch extractor of our own admonition.

When eyes (holes) must be punched in tool steel, problems can often occur. They can seem so insurmountable as to cause the smith to hang up the entire project. If punches and billet steel are ruined, what do you do? To drive an oval eye hole through a 3 inch round or square billet our initial punching may grab hold of the punch and not let go. When this happens, all are brought to this extractor apparatus. We do not have a name for this tool other than by function. The closest name that can be placed on this tool has been a stripper plate, for it strips forgings from lodged top tools. We are sure something of similar function must have existed in days of old, but none have yet to be discovered. For this implements invention, it was definitely form following the function as well as need.

Billets with lodged punches at any stage of hole development are lightly hammered on their ends and sides. This helps to elongate and stretch the steel surrounding the punch. The punch and billet are then brought to the extractor, if the two are still stuck together. All punches used in this extractor must have prominent shoulders at their upper (striking surface) region. This enables the punches to hang in our vises bolt on addition when scissored closed.

Picture 252

Shown is a 4 lb hammer head with a drift punch lodged within its eye, the shoulders of the punch hang in the "V" cut side of this apparatus. The vise is cranked closed to bring the opposing pointed side into the "V" cleft space. The point presses against our punch head and handle to hold it all secure.

Picture 253

The hammer head to be extracted is dangling from the lodged punch here. The point side will be cranked into our V cut space. The inverted "U" shape handled tool in the upper right now comes into work. It has a plate beneath which connects both ends of the "U" shape. In the plates middle is a cut out slot which permits passing around the lodged punch and resting above both striking surfaces of the lodged hammer head. The smith now has both hands free to hold the inverted "U" shaped tool handle and strike its top. The hammer head now just slides off the punch. Both sides of the hammer are pushed simultaneously and our punch holds still.

Picture 254
The wedged punch is closed into our extractor and
made ready for driving the billet away from the punch.

Picture 255
The hammer is striking our “U” shaped key and off comes the hammer head billet beneath.

THE HAMMER TOOLING, DIES

Picture 256

Pictured is a group known as set tools. The two wood handled tools at left, and four to the right come under the category of flatters. The wood handled flatters are made of S7 (shock resistant air hardening tool steel), forged, annealed, ground, hardened, then oven tempered. Their design is to be struck, lifted slightly and moved, then placed and struck for smoothing out imperfections and contours as needed. All have radiused working edges (not sharp) and crowned striking surfaces. This slight crown (highest in the center) to the striking surface, enables the smith and/or power hammer to hit straight through the set tool. Uneven billet surfaces are more easily managed with this feature. There is much less chance of cracking an edge or corner off of this design of set tool. The lower height tools are most suitable for trip hammer use.

The five set tools, pictured in center view, are truly set tools with specific radiused edges. They will not simply level the work of a narrow section, but leave the desired exit radius when the tool is removed. These tools are not for traveling about on a work piece, as is a flatter with gracefully smoothed and radiused working surface edges. These set tools have a specific width and length of surface they can cover to smooth.

Picture 257

At left are three set tools, the two at right are fullers. Note the differences of radius in their working surfaces. The handles remain nearest to the striking surface in each case to view the billet progress beneath. This group is made of unhardened sucker rod and have been in service for many years. These are used only on hot work-to-present-date.

Picture 258 (see opposite page)
Hole die, rails and top fuller side view.

Picture 259 (see opposite page)

Shown is a bottom hole punching die in the foreground. The hole was drilled completely through the die top to bottom. Punching is started on the flat unholed section. The punch tool is extracted from the billet. The billet turned over, and the hole punching process is continued through its blind (halo) centered side. As the punch is reaching our bottom die surface, it is gracefully moved to above the die block hole. In graceful movement of tooling and billet above this die an attribute is addressed. The operator is listening and feeling the upper die hit the hand held tooling to judge when all should be centered above the bottom die hole. The punch will then have its best opportunity to push out a billet hole slug or button remnant. Punching can now continue without mushrooming the punch tip or damaging the die block. The hole gives the punch a place to go.

The hole punch die is fairly low in height. Most often this die is used with top tools. Many mechanical trip hammers have vertical working height constraints - that height under which the machine will work with gravity, to rise and fall without grabbing our tools and throwing them around the shop. If we know what that vertical work height constraint is, we can operate beneath it comfortably. This height includes die height (upper and lower), billet thickness and top tool used.

The three rail tools shown at center are hole die mounts. When narrow sections of billets with large ends must be worked, we have found this to be a very good solution. These are used when the top and bottom die widths are too great. The desired width of rail is dropped into our holed bottom die. The whale tail/free end wraps against our left die edge snugly. It will now hold still during forging operations. These are made of remnant truck axles, unhardened after forging.

The top fuller shown at right is of the same width as its bottom rail counterpart. Forged from one piece, it too is made of unhardened truck axle. Scarring of the upper die is now minimized as this fuller is turned and rolled.

Many different diameters of fullers can be experimented with to get a desired result. The chief requirement is that the widths of the rail match that of the fuller while allowing clearance of its handle. A 2 inch diameter round fuller can only penetrate approximately 3/4 of an inch with its handle centered. The widths of rails shown range from 1-2 inches. The fuller shown is of 2 inches width and 2 inch diameter, the handle was forged to 28 inch length with a narrow throat. The handle will bend in case of miss hit and spare stinging the smith's hand.

Picture 258

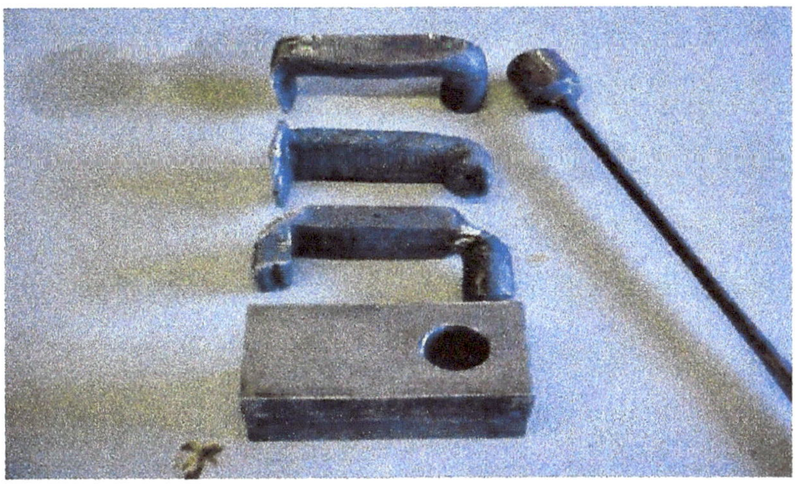

Picture 259

Pictures 260 and 261

Fullers can be made in all manners of shape, length, height, diameter and profiles. Handles as well are often varied by project requirements, and occasionally by shop materials available.

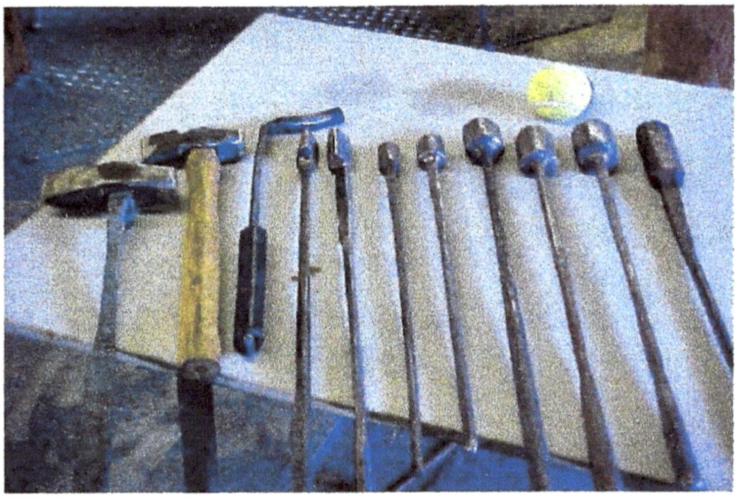

Picture 261

Pictures 262
The fullers to the left are for use under power hammer. The two at right can and are used under both hand and power. They have been heat treated for safety. All are used only for hot forging work.

Pictures 263 and 264
Set tools and flatters galore. These are used to form the jaws of box jaw and "V" jaw tong. Saddled bottom die power hammer mounts and top forming tools. The handles of some have been curved to enable working around the tong blank with an unobstructed view for the smith.

When tong blanks must be shaped all day, the correct position should be sought. Whether for the smith's body movement, or to achieve more exacting work results, a good view of forging in progress only helps!

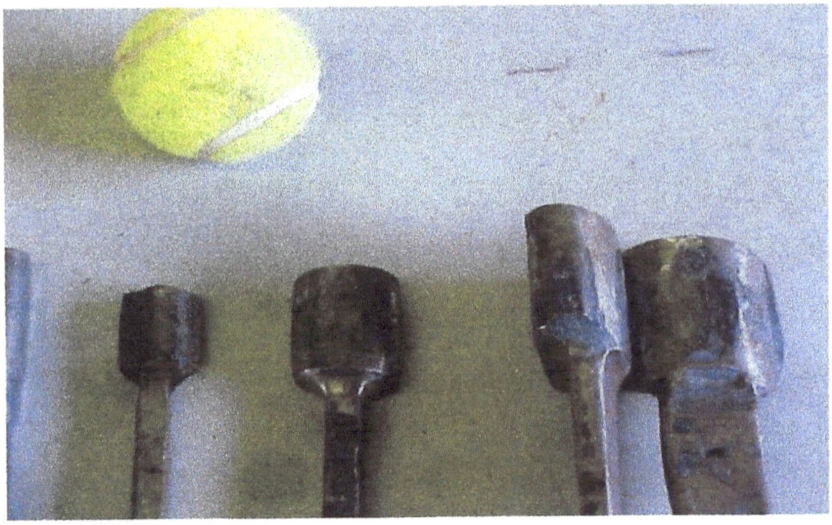

Picture 265
A group of flatters crowned primarily in width on their striking surfaces. Excellent for smoothing long tapered tools and forgings hot. Quarter inch radius on all working surface edges.

Picture 266
Such a variety of items – can the blacksmith toil to create and recreate? No two the same in shapc or purpose, yet all as a means to an end.

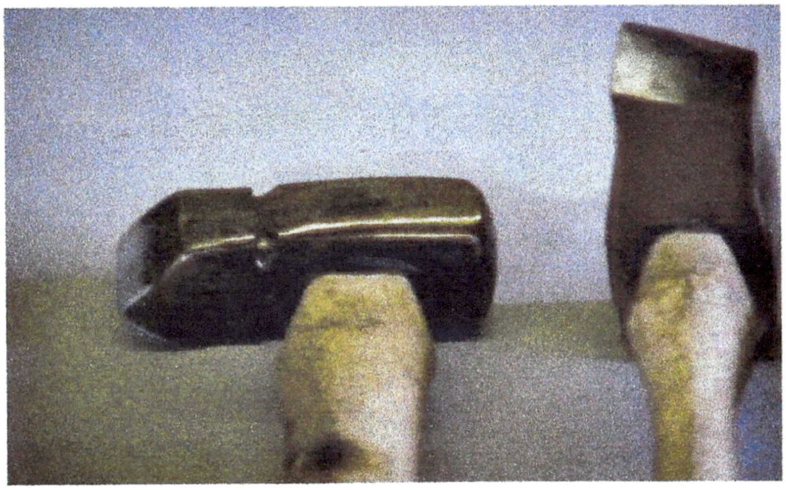

Picture 267

Convex "V" cut set tool of 90 degrees is shown at left. Used regularly to forge and form for tong jaws. A concave "V" bottom tool is used to back this tool up. Many other uses will be discovered when a smith has one of these.

At right is a side set tool. Used in our shop for maintaining one billet shoulder very acute and upright while setting down another. The handles are attached with bicycle inner tube. A 6 inch section pulled over the handle after rasping to fit. Then drive the hickory and tube through your tool eye until lightly snug. The excess rubber is trimmed away with a sharp knife.

Picture 268

To the left is a eye exit hole tool. For secure wooden handles, the tool eye should have an hourglass profile, smaller in the center and larger as handle eye exits the tool. The item shown is driven into a handle eye to make it broader and longer only on its perimeter; much like driving a football into a sewing thimble. It need not go far, but just enough. This is done one side at a time or with a bottom tool simultaneously.

The next tools are a good example of our chicken or the egg principle. Which comes first? In this case we made the convex forms, then drove them into our steel blocks to create concave dies. The concave dies did not receive handles until after all forging and heat treatment was complete. The mild steel handle attachment was done with DC negative reverse polarity arc welding. The rod used 312 stainless; no problems after years of use.

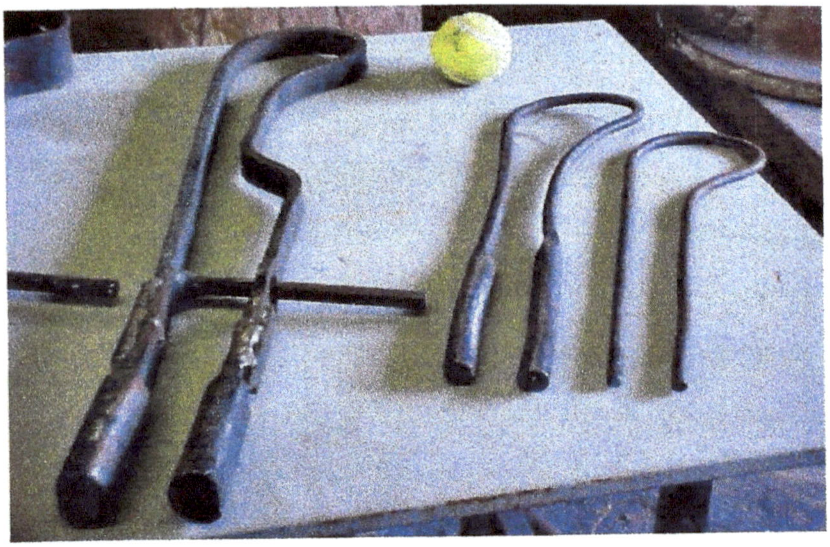

Picture 269
Hand held spring fullers.
The length, width and thickness of stock, radius circumference and shape in the curved section determines how stiff a hammer blow is required to activate the tooling. When reduced lateral deviation is a priority the spring section and handle is built wider and not thicker. When unobstruction of view and hand held comfort are issues the spring should droop to the lower most portion of the tool. Fingers are <u>not</u> allowed to drift between any portion of these tools!

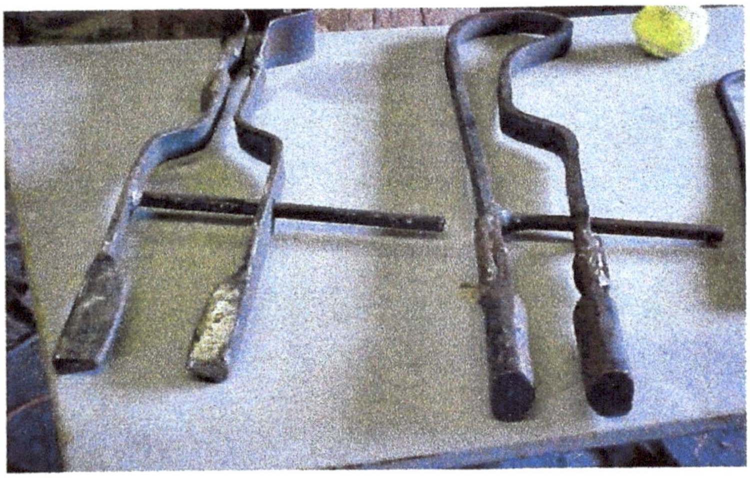

Picture 270
Where accurate work is desired, a pinned spring fuller helps keep each fullering surface opposing its counterpart. The through holes for pin passage have been hot punched to preserve bar stock mass rather than drilling.

Picture 271
Hardy hole mount spring fullers and swedges used under trip hammer power. A square hole saddle type mounted to our bottom die makes for easy change out of these important tools. Note that each has a different working surface, yet their individual working heights are virtually identical. They are all set up to operate against a 3 inch round billet.

Picture 272

One of the shop shelves is designated to supporting trip hammer tooling which is in regular use. These items are installed and removed on a regular (daily) basis for our various tool making operations. It has been said that tools must be made to make tools to make other tools which can complete a project. Here are but a few made which stand by ready to take on final project elements under a power hammer. They are designed to be hefty enough yet efficient under a 100 lb head weight.

The dies to the front right are of combination set up for drawing length or width and flattening. They measure 3 inches wide x 6inches long. The heights vary from 3-1/2 inches to 4-1/2 inches. The drawing (half round) surfaces are most often installed in the power hammer to oppose each other. They are made of 4130 steel from 8 inches round train axles cut and machined down with dove tails for taper die key keepers. The old school attachment of dies is a taper key and still works well for quick set up, tear down and alignment. The dies were water hardened after annealing, deburred, and tempered to brown color.

Each die has its own single or pair of taper keys. Those keys stay with their intended dies whether in or out of the power hammer machine. This makes for a snug fit time after time. Chipping or scarred keys are redressed or replaced. Some are saved to be used for a future die set. Most all of ours are made of unhardened truck axle steel. These are tough, yet not hard enough to scar the die or machine when hand hammered into position. If hardening the keys is desired, anneal first; oil harden from the critical (non-magnetic) temperature and then temper to purple.

Two of our taper keys are shown next to the flat die ready for use.

To the right rear are spring fullers set up as saddle mount types. A mild steel plate with sides which straddle a flat bottom die. These enable top and bottom narrow fullering with the smith's hands free to move other tools and billets around. The fullering surfaces and spring portion are all made from one piece of sucker rod steel one inch round. The free upper fuller striking surface is crowned in both length and width. This enables the power hammer upper die to find the spring fullers center continually as the forging it creates becomes narrower.

To the left are all saddle mounted bottom swedges. Their chief importance being the depth with which they allow material to be worked. The collar and hole depth was decided upon for power hammer efficiency; that height at which the machine develops good rise and fall with gravity and moves our billet steel without excessive force. These particular ones shown are used to set down the working surface edges of various flatters. On occasion, the wrist pin collar section mounts have worked well for use as measuring gauges and stops. When running off multiple production items, which must be identical, bar stock is laid across the span of these collars and hammered leaving two indentations. The bar stock is then picked up and moved toward the trip hammer. One indentation is now hanging in mid-air, the second notch on the upright collar side, farthest from the smith. This continues until all is marked as is desired. All notches and indentations are then equidistant.

Picture 272
The every day die shelf.

Picture 273
Known as a combination flat and tapered bottom die. The height of this die enables working very narrow pieces of billet, yet not crashing the upper die to its working surfaces. It has been used to put the final form on many claw hammer claws. The upper die which opposes the tapered side is flat (no taper).

Picture274
Top and bottom spring fullers, flatter, and swedge for hardy (square hole) mounting. These are used in a saddle mounted square hole beside our trip hammer bottom die. The spring and working surfaces were made of one piece of 1 inch round sucker rod 2 feet long. Forged to form and not heat treated. The square bases are of remnant truck axles and arc welded for attachment. (Same as Picture 271)

Picture 275
This is a bottom die and saddle for a 100 lb trip hammer. The through hole is used for punching eyes for other hand tools to be used in the shop. Height adjustments are easily made by adding or removing the saddle to accommodate vertical workspace requirements during forging. For specialty dies, the die taper keys are kept with the intended die. This keeps change out of die time minimal and is sure to hold when in the machine.

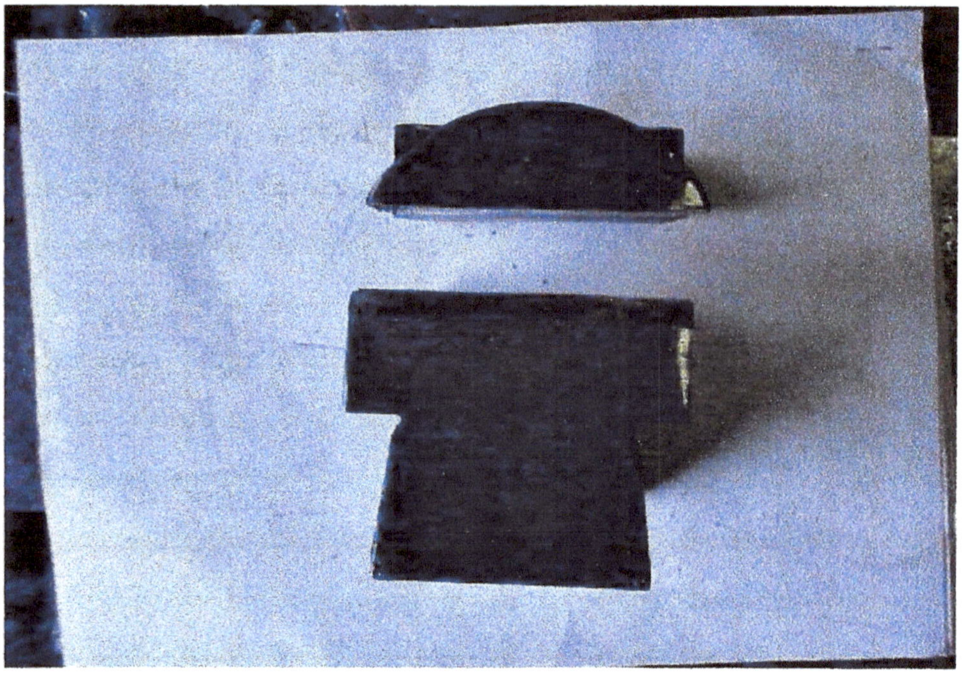

Picture 276

A combination flat and half round convex drawing die. The top is being made to attach to its flat dove tailed bottom.

Picture 277

A combo die in our box leg vise. Heat treated before final arc welding attachment. A top and bottom pair of these were made for our 100 lb head weight trip hammer. A great asset for a powered blacksmith shop. Made of 4130 train axle.

Picture 278

Three types of heading dies for use under trip hammer power. The bases are set up as a saddle to surround the lower flat die of a 100 lb trip hammer. They differ in diameter by 3/8 inch increments from smallest to largest. A tennis ball is present for size reference. These are all forged dies made of spring steel, remnants of 2 inch diameter 5160, forged to square. A positive (male) form of the appropriate diameter is made first of a carbon steel. Truck/car axles, other spring steel, new air hardened steel or something handy that can be hardened. The positive side is forged then annealed, ground and polished to size, hardened, polished, tempered. Lastly arc/mig weld a handle on, if one was not forged out already, it must be made first, then attach it to the positive tool. The tool handle can also be forged out of a portion of the positive billet. All but air hardened steel works very well for the positive tool for our operation. As the positive tool is in use – being driven into the hot steel which will become the bottom (negative) die, it will pick up heat. As the forging heat diminishes, this positive tool can be quenched out in water to cool and be made ready for the next negative die forging. If it is made of air hardening steel, the best remedy may be to set it aside to air cool and have a cold one (spare) ready and waiting - another reason why we mark each tool handle with the type of steel the tool was made of. Quenching the air hardening steel out in water/oil over time may cause it to fracture. If immediate cracks are not present or visible, they may show their wrath when the tool is in use. Flying chunks of fractured steel can and will do harm. Even oil hardened steel quenched in water to cool may not be good. For when the tool is in use may develop a fracture. The best medicine for this is to not over use the positive tool during each forging of your negative impression. When the positive starts warming up stop, put it down. Toss the negative block back into the forge, then quench out the positive die. For most all of the positive (male) die makers through forging, there is an appropriate operating time before they become too warm to use without damaging their form. A hot positive form will then loose its crisp edges and overall dimension.

(Picture 278 – continued)

The negative heading dies shown are for farrier and blacksmith tong jaws. The slots which lead away from the bowl of negative dies were made with a slot punch, wide enough to reach from mid bowl to beyond the outer perimeter of our billet block. Both top tools were used alternatively until the desired depth was reached. After each top tools use, the block is rotated and forged by itself to create a proportion back to near square. The block is reheated to yellow and again top tool forging is commenced into the block. When all forging both with top tool and the block by itself is completed, wire brush any free slag out and view the results. If all appears well, reheat to a dull red and anneal.

After annealing is completed, its time to dress the die to near finished. The tong neck slot was cleaned out with an angle grinder (36 grit wheel) with rounded edges; then to a flapper disc in the angle grinder (60 grit).

The contouring and removal of rough slag is the goal here. Much shouldn't be present if the forging stages were clean. Some will say its not how deep you plow, but rather how long you stay in the field. We are just cleaning and contouring. All external surfaces should be roughed with the heavy angle grinder (36 grit) if available; then smoothed off with a belt sander/grinder to (60 or 80 grit) finding the true edges where each internal and external surfaces meet. Rounding the crown of these edges may be necessary to prevent fractures during the hardening process. They can later be sharpened as well.

Time to heat treat, or do any special die grinding if desired. For spring steel or axle steel, bring the die slowly (10 minutes to one hour) up to critical heat. We call the critical heat that temperature where the steel starts to loose its ability to attract a magnet. So have one handy by the forge to test this continually, until your eyes can find this color on their own. The heat up time is billet size and intricacy dependent. A die weighing one pound will take less time to heat thoroughly and uniformly through its center than a four pounder. Heat up time is shape dependent as well, so rotating your die blanks in a coal forge is a must. Quench out in warm oil or warm salt brine to cold. The next step should be to temper; commence immediately.

To temper: Oil – hold to the fire and flash off the residual surface oil, then allow to air cool.

Salt brine – wash off residual salt with fresh water. Grind the external surfaces to clean which enables viewing the temper colors. Reheat to draw purple-blue color. Allow to air cool if small, (6 oz.- 1 lb.) If larger die (1-4 lbs.), quenching in warm water (tepid) may be necessary as the residual heat present can continue the tempering to a color which is undesirable for the dies longetivity (grey) and render it too soft.

Picture 279

These are all saddle type bottom die covers for a 100 pound trip hammer. Pictured left are tong jaw heading dies arc welded to the saddles. The welding rod used was 7018 MR for attachment. Using 312 stainless is an excellent choice for joining two dissimilar ferrous metals. The die is 5160 spring steel and the base plate of the saddle is A36 mild steel. The 312 stainless is used for this as a preference but quite a bit more expensive.

Attaching the die to the saddle is the last part of its construction. The positioning of the heading die is crucial to ensure our upper flat die edges impact where desired for a given forging. Hefty cold shunts and defective forging will result if the upper die's near edge impacts the outside of the heading die bowl. This is also very dependent on the shape and volume of steel which is previously forged to be set down into this type of header. Numerous forging runs are often needed to find the correct volume of blank which is required to fill a die and appropriate shape of blank which provides a shuntless end product. Overfilling a die of this type has proven acceptable as the excess can be shorn or trimmed off with another trimming die. Small amounts can also be ground off by hand. Yet the closer to the ideal volume in your blank, the less waste will occur and hence improve production cost considerations.

Top right is a saddle set up cutting plate. The plate is made entirely of A36 mild steel. As the hacks and hot cutting chisels finish a cut, the bottom die is in grave danger of being nicked. Even with the best running machine and a good smith, misjudgment of force and thickness of material can occur. The color of heat in that material can dissipate or regenerate from the impact heat of friction (pounding). This plate can and will save a tool from having its edges marred or mushroomed. Expensive bottom dies are also protected. This plate is deemed sacrificial and is made of only mild steel 5/16 inch thick.

In the right middle is a 3/4 inch plate of structural steel. Mild steel of 3 /8x1 inch are welded to its lower edges to form its saddle operating function on top of a bottom flat die.

The function of this particular saddle plate is to aid in punching rivet holes in tongs. Note how close to the far edge these holes are positioned. The rivet holes are started on a flat section of the saddle with the tong jaw hanging off of the far edge. The round rivet hole punch is extracted and the tong blank rolled over 180 degrees. Our hole punched is then positioned over the halo left from the original punch driving. Work is still being done on a flat (holeless) section of this saddle. The rivet hole punch is again driven down to bottom out just lightly. Now our project is moved over (slide sideways) to be positioned over the correct saddle hole. Now the rivet hole button is driven out into the saddle with a crisp tap from the power hammer. This saddle serves much as a pritchel hole in your anvil. The size and position can be altered to fit the smith's needs. For best tong leverage, the rivet and rivet hole needs to be as close to the jaw as possible. Many anvils have a set distance from far edge to pritchel hole. Making a saddle such as this to rest on an anvil or trip hammer bottom die means we can create any distance imaginable.

Pictured at bottom right are square carbon steel rails arc welded to a mild steel saddle plate. The upper edges are rounded, not sharp. This saddle is used for setting down box jaw tongs of a specific width. Having a gentle upper edge on these rails for the initial set down of a flat jaw tong blank, it can become a box jaw without the external scarring that sharp edges would impart. A square set tool is held atop the tong blank which will allow for twice the blanks thickness to fit between the rails. If a closer tolerance is then desired, the same set tool is used in the box jaw tong blank but it is set into rails of closer tolerance to the size for which the tong will need to hold. Your distance between these rails can provide for infinite combinations. The combination being the width and length of set top tool and width and length of bottom rails to oppose the tong jaw blank. These rail combinations are set up to accept other materials for artistic blacksmithing, decorative as opposed to functional endeavors.

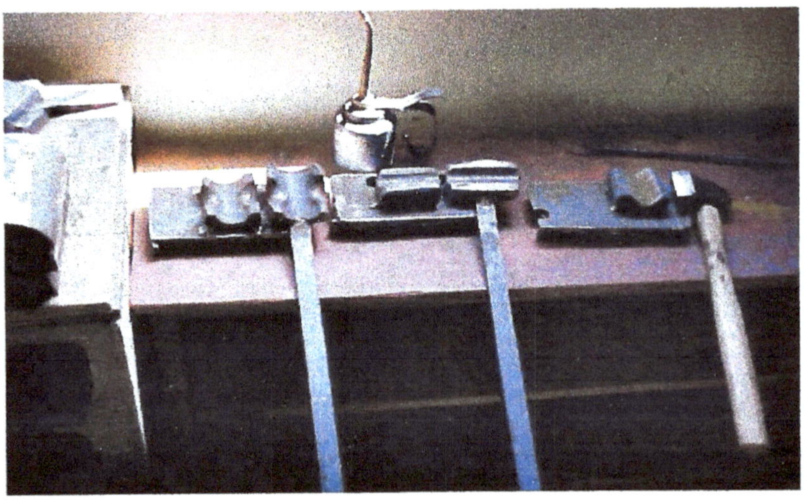

Picture 280

At left of the pump oil can is a top and bottom swedge used to contour the eye sides of 2 lb rounding hammers. All pictured are set up as saddle bottom die mounts for trip hammer operation. They will work for air and steam power hammer applications as well. The chief requirement is only to have good flat top and bottom dies installed first.

The middle (Picture 280) shows a top and bottom ½ inch round swedge. This particular swedge combination has both tools raised in height on their right sides. It is used to taper and smooth a forging which will gradually taper from 1/2 to 3/8 inch round.

At right is pictured a top and bottom *"V"* jaw swedge for hollow bit tongs. All pictured enable good and efficient use for a one man operation.

Picture 281

In the foreground are pictured trip hammer dies, both upper and lower. They measure 3 inches in width, 6 inches long. The height of these dies are all variable. One of our chief considerations in operating a power machine with these dies is finding a comfortable and efficient working height, both for the machine safety as well as our own. The vertical working space is of paramount importance to provide optimum impact imparted to the forging work and the tools placed between your upper and lower dies. These dies operate in a gravity dependent machine. The upper die merely rises and falls to deliver its useful force to create a given forging. To optimize the ability of our machine, use gravity and compensate for the changing thickness of a forging, the dies are changed periodically to achieve the most appropriate vertical workspace.

To consider workspace, suppose the trip hammer will function comfortably against our work at a range from zero (upper die down) up to 6 inches (upper die up). This is our total height of working capacity where the machine is not breaking its own parts nor throwing tools out of the operator's hands from unreasonable compression. The working height includes both our billet to be forged and any top or bottom tools in use. For example: a billet of 3 inches plus any tooling of 3 inches in height reaches the maximum working vertical height acceptable. For us finding this ideal height is one of the first chores sought after when running new dies on any machine. Adjustments are then made in the die height, making taller or shorter ones to find an ideal for our project at hand. For us, this die height change out has been the quickest and most sure machine adjustment during a days shop work. Re-adjusting machine armatures, toggle links, tension spring, mechanical brake assembly, etc. will take time. Reference marks on the machine are needed to remember where each was set for a project. Simply changing out a die once in a while after all other moving parts have been set and tightened should take only minutes from the work day. Periodic checking and oiling is then the only requirement.

The dies pictured are all made of 4130 steel hardened in warm salt brine. They were then tempered back to straw on the drawing dies (half round). Dark brown – blue on the flat dies and combination flat and drawing dies.

The steel used was a remnant of salvaged railroad train axle which began as 8 inches round. In the annealed stage, it was band saw cut into blocks and all dove tails were then machined in. The upper one inch of the tallest combination and drawing dies is a separate slab which was also machined before attachment. All heat treatment is done before attachment of the upper slab to the base block and dove tail. As the upper slab (also 4130 steel) is thinner than our die base blocks, there is considerable chance of warpage when it is attached to the base before heat treatment. Different thicknesses cool at different rates. The upper slab is attached with a DC reverse polarity arc weld of 312 stainless rod. Rod such as 7018 and others will suffice, yet all should be clamped and tacked together on all corners before heavy beads of weld are laid down. Near complete contact of the slab to the die block is needed to prevent the welds from being flexed and later work hardened to crack during die use.

The attachment of each die on our old machines is made with die keys. A long tapered wedge similar to that for firewood log splitting.

The keys, as shown, are kept with their respective die which fits to the machine snugly. When the keys are driven into the machine to cinch each die, their thickness and length of taper provides for protrusion of both the thicker end which is struck to tighten and the thin end which exits the dove tail slot. This makes for ease of removal, though should not be excessive for operator safety. The keys were made of remnant truck axle steel not heat treated; simply air cooled after their last forging heat. These can be heat treated if desired in oil and tempered back to blue-purple. They should remain softer than the hammer head or base frame they rest against. They are more easily replaced than ruined machine dove tails, and therefore need to be the softer and hence the sacrificial part – if you will.

To the rear of our picture are saddle mounted spring top and bottom fullers. The working part and spring were forged of one piece of sucker rod. They are used for necking down narrow sections of forged material.

To the right rear is a large hole punch saddle and matching hole die beneath. Variable height compensation is easily made with this type of set up for drifting (enlarging) small holes to larger ones. The hole punches themselves can be switched from shorter and thinner types to taller and thicker types without changing out the complete bottom die. Where working height becomes dangerous and inefficient for optimum impact, from our upper die using larger hole punches, the saddle is removed to increase our vertical workspace.

Photo 282

The implement captioned here is called our set fuller. This tool is double hinged and spring loaded for tensioning. This one is saddle mounted to the lower trip hammer die. The fullering rods upper and lower are 5/8 inch diameter and 10 inches long. The double hinge aspect allows for fullering material from 4-1/2 inches thick down to virtually nil without making tool adjustments. The upper rod is hinged to compensate for height of material changes as work proceeds. It maintains a parallel position to the upper trip hammer die striking surface through all material thickness changes. The fullering rods are made from good quality air hardening tool steel. All else is from mild steel (A36) and sucker rod to forge the upper moving hinge rein. The tensioning spring was salvaged from an old trampoline. A "U" clamp, brace, lock washers and nuts enable the most appropriate spring tension to be found. Once found, the tension has not needed to be readjusted. The spring delivers just enough down pressure to keep our upper fullering rod against the work during forging. This has been an indispensable tool for it allows the smith's hands to be freer and his eyes can focus more on the billet progress.

Picture 283

Front view of our set fuller for trip hammer use.

Picture 284

The double hinged set fuller in position for use. It rests on the lower flat die. The vertical guide to the right helps keep the fullering rods aligned with each other. The guide is calibrated in half inch increments to help find the depth of progress which has been achieved without a work pause. The weight of the guide also acts as a counter balance for our open and close reins. This particular tool is only resting on the lower die, not clamped or bolted and enables for a quicker change out of tooling.

Picture 285
Side view of set fuller in the ready to use position. It was waiting to fuller down 2 inch round head billets this day.

Because fullering is one of the most dramatic constituents a smith can add to their work, some effort should be made to try many means of its accomplishment.

Picture 286

Picture 287

Pictures 286 and 287 to the left.

Pictured are all different cutting implements for use under trip, air or steam hammers. They will also function efficiently under mechanical hammers such as treadles and olivers (foot operated, gravity driven striking hammers).

Each has an overall length of 3 feet and have thinned mild steel handle throats for safety. In the event of misplacement or missed hits, this type of handle will bend rather than transfer all shock to the operators hand, wrist or arm. The tennis ball is present for size of tools reference.

All tools pictured are used for cutting and marking hot tool and mild steel or non-ferrous metals, such as brass and copper. The working ends (heads) of these are made of air hardening tool steel (S 7) hardened and tempered to straw. A small 1 inch tang was forged away from each head for gas welding (oxy/acetylene) the head to each handle. A scarfed forge fire jump weld can be used for the handle attachment as well but much attention should be paid. The tool steel can be damaged even with the utmost of attention.

The four tools are hot chisels with acute (sharp tapered) working edges. The offset handles are a good addition when the smith's field of vision cannot be compromised.

The four tools to the right are called "hacks". They are ideal for cutting larger (1 inch thick or greater) round and square stock. Their working edges are narrow but blunt (not pointed). The profile of the cutting edge provides for two cutting corners which shears through the material being worked, thus the term "hack" is applied.

A taper exists from the top which is stricken to the edge which does the cutting. These tools are kept sharp with our belt sander when the edges become battered or rounded from the heat of use in cutting.

See original diagram-next page

Hacks can have cutting edges which remain parallel to the material to be trimmed or carry an angled profile. Our preference on the striking surface is to keep them slightly crowned (higher in the middle) when looked at from a side view. This is also true when viewed endwise.

The upper (striking) flat die can now drive the hack with less trauma to the smith's hands. Finding the appropriate width, height and length of hack for a project is a good idea.

For each project there is an appropriate ratio of hack size (mass) to working billet. The hack will pick up radiant heat from touching and/or remaining near the billet it is cutting. Therefore, must be large enough to accept some heat before the cutting edges show color and fail (mushroom). The hack also cannot be so large that pushing it through a hot billet is tedious or cumbersome, requiring excessive force with minimal progress of the cut. Experimentation with various hack dimensions should be challenged. In many old shops that used hot cutting tool methods, a hack may be hard to find. They were used and redressed until quite literally, no tool was left to cut with.

As a general rule, we keep two or three hacks of near identical dimensions near each other. When cutting a billet of 3 inches round stock, even the ideal hack will become heated and red at the cutting edges. This heat will render the sharp edges very susceptible to rounding and becoming mushroomed. Don't' wait for the color of heat to show. As the first hack warms up, put it down on the nearby steel work table and pick up the next cool hack of the same size. The billet is rotated to allow all sides of the cut to be accessed. The last of the cut for this steel dimension is through its center, leaving only a small ribbon to be hot chisel trimmed if needed.

To the far right is saddled mild steel cutting plate. It is placed over our buttom die to prevent damaging the die or cutting tool as it finishes a cut.

Becoming familiar and handy with these cutting tools can make purchasing very expensive power machinery unnecessary. Learn their attributes, both capabilities and limitations carefully.

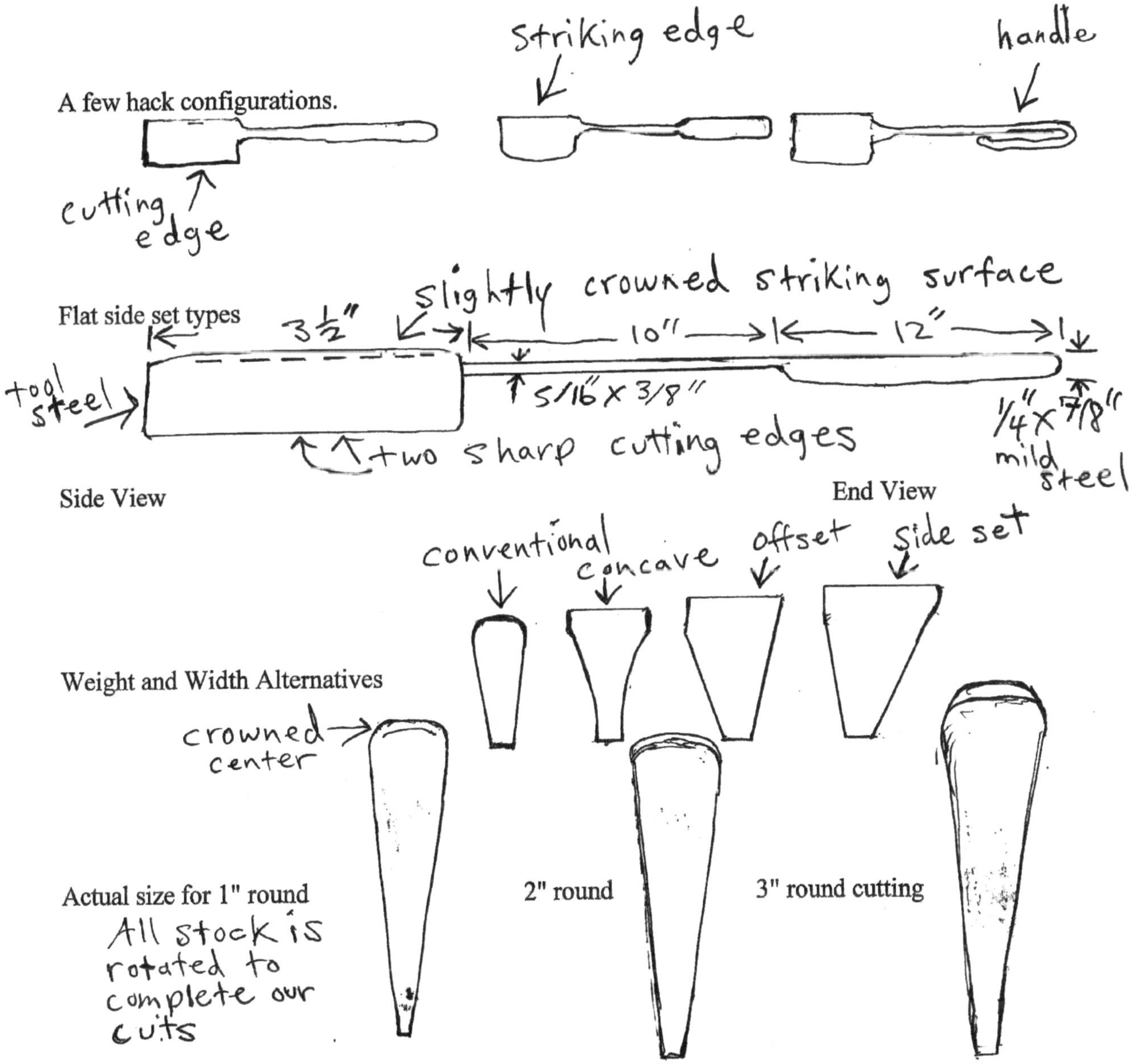

THE FARRIERS PULL OFF, HOOF NIPPER, AND NAIL CUTTER

Picture 288 to the left.

To the left we have a farrier's pull off tool, used to remove old horse shoes and nails from horses' hooves. The knobs forged on the round rein ends help distinguish which tool is which, when laying in a work tool box. Made from a 1-1/4 inch diameter hexagonal jackhammer bit. The length cut for each half was 5-1/2 inches. Upsetting was not performed on any portion of these forgings as should be the case with most types of tool steel. Wider portions have been flattened and narrow portions have been drawn. The chief principle is to start with a diameter of steel large enough to accommodate the widest and thickest requirement needed. Then leave a little material during forging for finish grinding and dressing. The amount left will vary from one smith to the next whether achieved under trip hammer or through hand hammering. This is where the individual must develop an eye for perfection!

Picture 251 to the right.

The tool to the right is known as a nipper. It is used for trimming the hooves of horses, cows, sheep, goats, etc. The shoulders of its jaws and reins have been left massive to provide strength to cut through some of the most dense and durable hoof material. The only limiting factor will be the strength of farrier's hands to close the jaws through a given hoof. The reins have been swedged to half round (convex side out) to allow for comfort and wider surface area when in use. No knobs at rein ends were forged to distinguish from the pull offs. These too are of jack hammer bit steel. They will be hardened in oil after a slow rising heat to a non-magnetic temperature.

The jaws to mid rein are hardened, moving the tool up and down in the oil quench so as not to render a fracture point in the reins where they emerge from the oil surface. The tool halves are then tempered immediately. If left in the hardened state for even a day or two, surface cracks or worse may develop. If old crank case oil was used to harden, the excess oil is allowed to drip back into the holding tank or bucket. The residual oil left on the steel surface is used as a gauge for our tempering process. While still slightly oil coated, the tool is presented to but not in the forge fire and rotated by a separate hand held tong. The oil will begin to exhibit a flash fire dance of its own. Good air ventilation must be ever present as this smoke will have an unbreathable toxicity. Fans and mouth breathing respirators are helpful along with an open air environment. Closed tight shops without overhead powered air ducts are unacceptable!

As the oil flash subsides, the tool is removed from the flame and is termed tempered. The oil is not doing the tempering, it is but a gauge to help our eyes recognize what temperature of heat has been imparted to the tool. If the tool is to be hardened in non- flashing oil, most surfaces should be ground and polished before heat treatment.

After the non-magnetic temperature is achieved and the tool is quenched to cold for hardening, the residual oil is wiped off. The tool is inspected to view clean chrome surface steel. The temper colors will be drawn by the forge heat, and we need clean polished/ground surfaces to view what these colors are doing. They need not be perfect surfaces, only visibly clean to see our oxidation /reduction temper colors. If the smith spends too much time re-prepping the steel surfaces, cracks may develop, for the tool is still in the hardened stage. A good bench grinder wheel for fast surface cleaning is the deburring; medium or fine grit.

This helps remove any burned oil and scale fragments. With chrome appearing surfaces, we can now temper the tool. The tool is presented to the fire and slowly rotated. Our goal is to have the jaws show brown-straw color and the reins purple-blue. Heating the rein shoulders first will help achieve this as they are thicker and will need more fire time to heat. The heat of transfer here (thicker portions to thinner) will be our friend. Doing this slowly, we will not need to dip the tool back into a quench (oil or water) medium. They are allowed to air cool after the appropriate color is achieved.

If a coal fire is used, the smith has another variable to contend with. The coal smoke will shade and grey the steel surface making color detection very suttle and disguised. A mild steel plate should be placed over the established coal fires and warmed to cherry red. This will act as both a heat sink radiator and smoke shield for the tool. The plate should be large enough to cover the entire fire pot and/or shade the entire tool from coal smoke. The plates thickness is usually 1/4 inch for most of our hand held tool tempering. A gas fire does not have this dilemma.

Both of these tools will cut horse shoeing nails without adverse jaw blade damage when hardened and tempered correctly.

The pull off tool (upper left picture) can be made to cut horse nails and small construction (building) nails all day long by leaving the jaws more substantial in their final dressing.

Jaw blade lateral view profiles: original diagrams

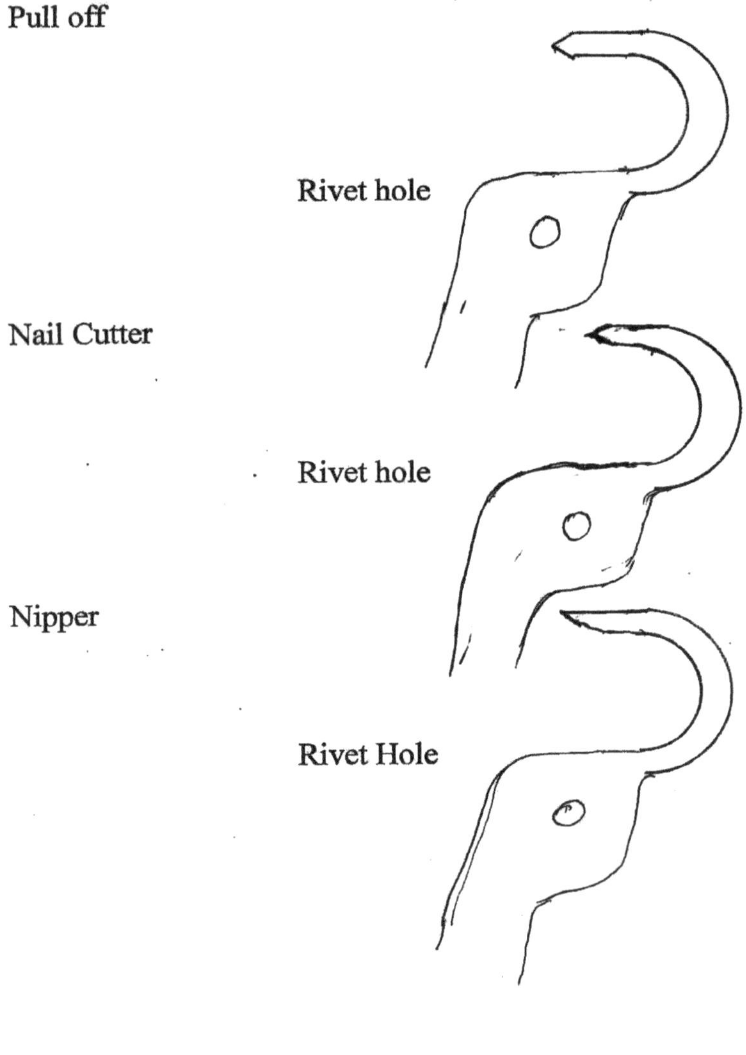

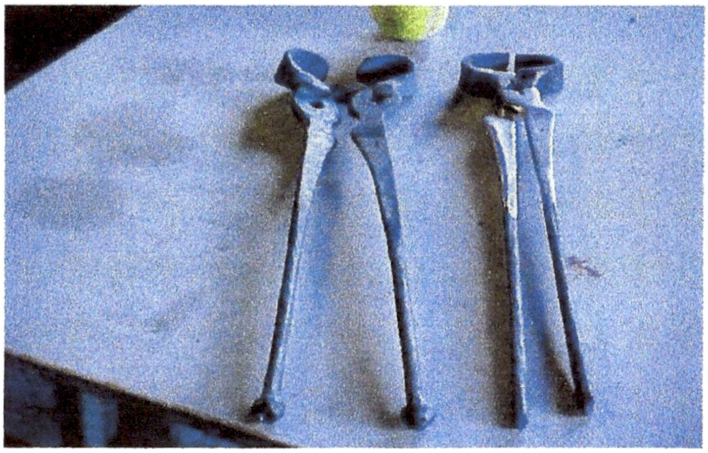

Picture 288
Farriers pull off;
hand forged at left.

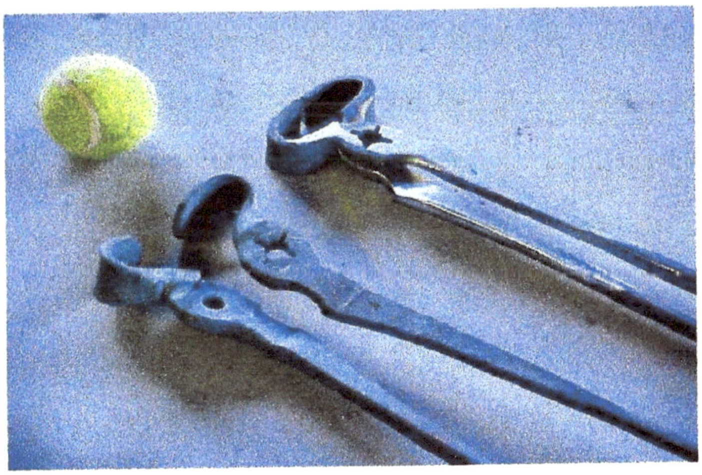

Picture 289
Hoof nipper at right;
power hammer forged.

Hand Forging A Farriers Pull Off Tool

Progressive forging sequence; pull off, nail cutter, hoof nipper

Appropriate materials: tool - Jack hammer bit, coil spring (5160), sucker rod, tractor coil (implement)

rivet - mild steel (A36)

Heat treat (post forging): anneal, harden (water or oil or salt brine) steel and use dependent; rein shoulders → jaw only -

Temper: straw - brown - purple - blue steel and use dependent

All ok. for shape to start

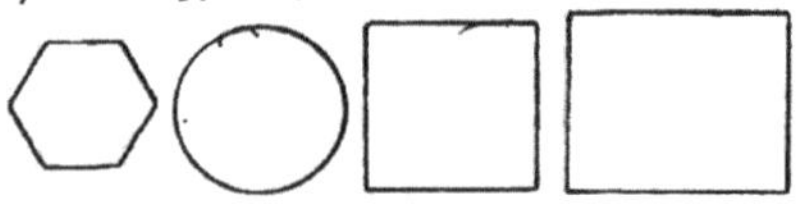

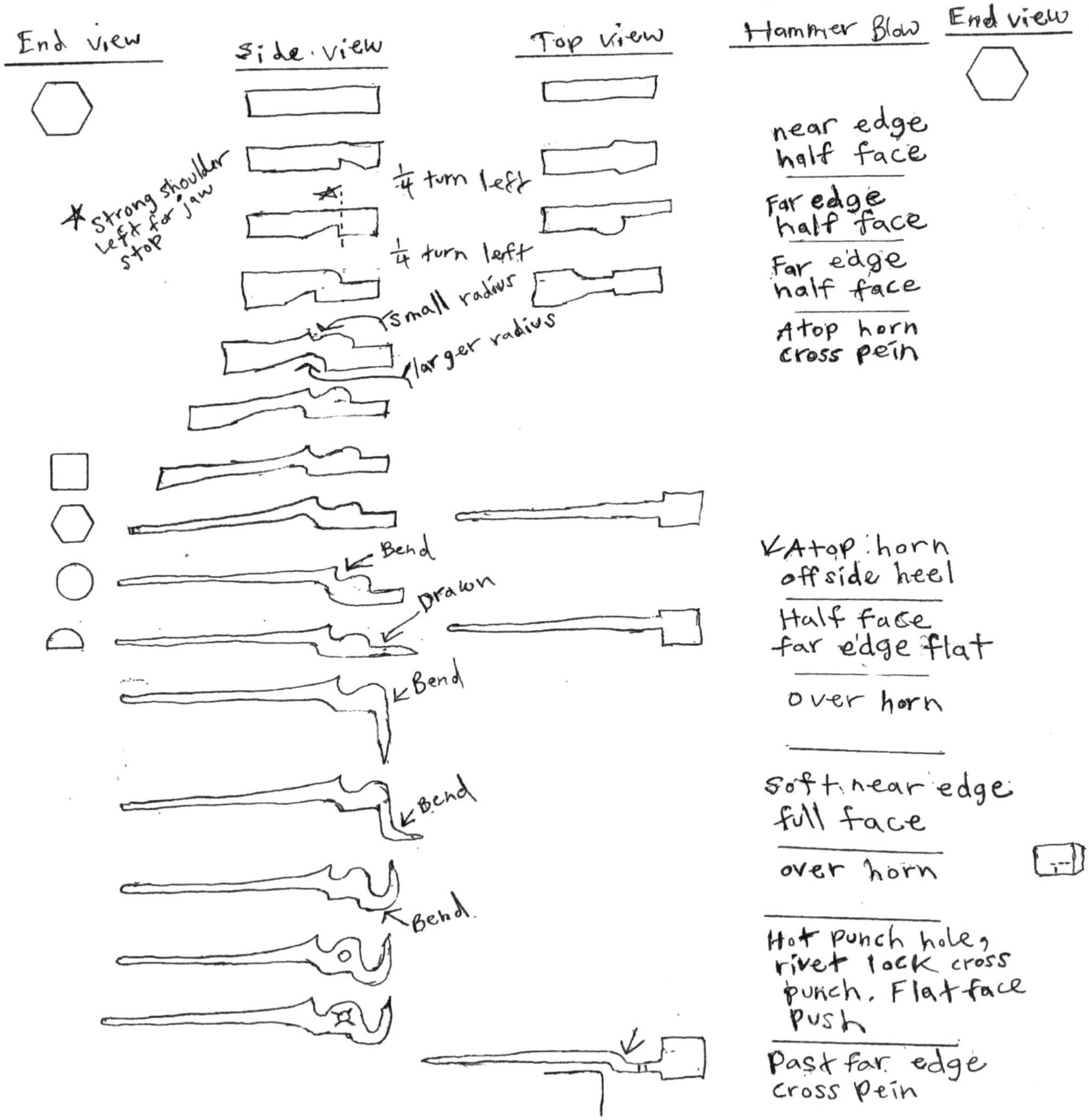

BY HAND OR MACHINE?

Picture 290

View of *"V"* hardy shank (portion which enters anvil hardy hole). All of our hand made hardies are marked on their shank. Because all of our shop anvils were also hand forged from days gone by, the square hole is not perfectly square. As we forge new hardies for each anvil, they are driven in that anvil to fit most securely, and fit its unique voids. After that hardy blank is *first* extracted from the hole, each is given a mark to the anvil horn shank side. Now when the blank is set back into the hole for future forging and later project use, we can always insert it as the forging will best fit. The mark here never lies.

Picture 291

A hardy used to make box jaw tongs. The tong jaw must only be wide enough to bridge the center of the hardy.

Picture 292
Box jaw hardy bridged by a tong jaw blank. The tong jaw only needs heating and we're ready to set it down.

Picture 293
Box jaw hardy in the hardy hole, tong jaw blank above it, flat set tool above all. This set tool will fit between the upright posts of the hardy including the thickness of the tong blank when driven downward.

Picture 294
The tool shown serves the same purpose as the posts of the box jaw hardy above. These two versions exemplify the exact point where hand use only tools meet their counterpart for power driven machinery. This plate is straddling the lower die of a power hammer. The plate itself and perimeter are of mild steel. The set up is often referred to as the saddle design method. The rails visible are of tool steel and can be welded on at any width apart for the smith's needs. They serve as the uprights between which the tong jaw blank is bridged.

Picture 292

Picture 293

Picture 294

Picture 295
The power hammer box jaw rail saddle with tong blank in place for forging and forming.

Picture 296
The same box jaw blank with a top set tool in place. The power hammer is ready to set the jaw down between the saddle rails. The width of the set tool must allow for the thickness of our tong blank jaw. The set tool handle is arranged to run over and behind the dies. This provides the operator an unobstructed view of the work at hand.

SET TOOLS USED BOTH BY HAND AND/OR UNDER POWER HAMMER

Picture 297
A top set tool for swedging the crowned tong jaws. The cleft, offset beneath the wooden handle is to allow for throat clean out leading to our rivet boss (tong bearing surface).

Picture 298
A specialized set tool for dressing half round rivet heads. The center is dimpled to accept an appropriate head volume of steel present on most of our 3/8 and ½ inch diameter rivets.

Picture 299
A *"V"* hardy with softened rounded upper edges used to form many tong jaws. Made of 5160 coil spring steel hardened and tempered with oil.

Picture 300

The convex *"V"* cut top tool with a wooden handle is ready to set a tong blank down. A concave "*V*" bottom hardy is beneath. The profile of both *"V"* tools is identical (90 degrees angle). The bottom tool has softened upper *"V"* edges to permit steel to enter without excessive scarring.

The top tool does not have to be the same length as the entire tong blank it will form. If it is shorter than the blank, simply start forming the fold away from you and draw the top tool towards you, walking the tool along with overlapping placement. This shorter tool does not need to be struck relatively hard to have good results. Keep the tong blank centered above the hardy.

Picture 301

Hollow bit tong blank: the flat jaw is resting above a *"V"* cut hardy (anvil square hole tool). It will be driven into the bottom tool with a convex *"V"* top tool to become a *"V"* jaw. The top is a set tool merely rested on the hot tong blank and then itself struck with another heavier hammer. This same top tool and saddle mounted bottom concave *"V"* swedge is powered through our trip hammer operations.

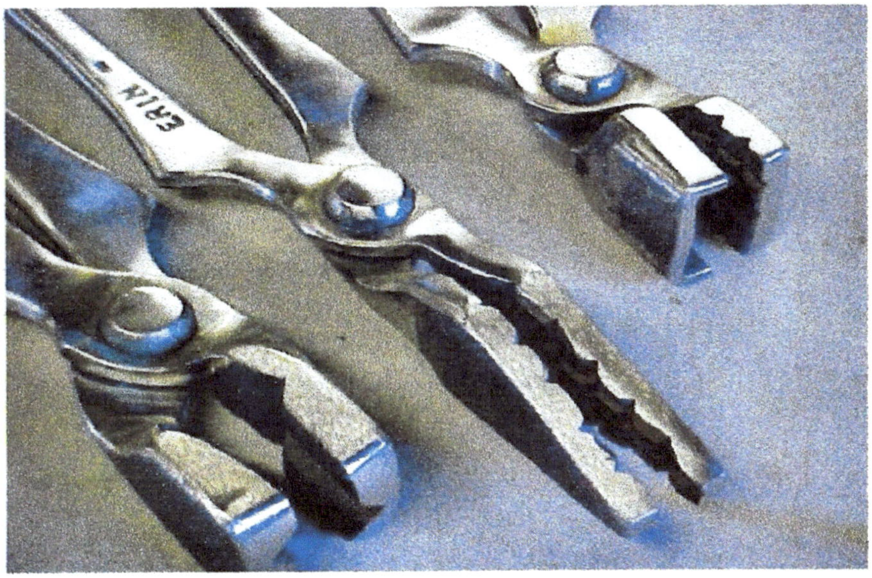

Picture 302

Left to right: Our flat jaw tong with *"V"* cut length wise and cross-wise through working surfaces. This one will hold 3/4 inch square and round stock very securely.

A wide mouth alligator jaw tong has been cross hatched with a hot chisel for added grip and holding security. These are great for holding brazing rods and bending hot steel in tight quarters.

A box jaw tong to hold 1/2 x 1 inch flat stock, cross hatched flat jaw side. The same amount of steel was appropriated for all of the jaws pictured; from tip of their lips to rivet. Such a wondrous variety forging can enable.

The only specialized tools used to create all of these tongs include a hot chisel, a square set hammer, a "U" bottom tools, and a *"V"* top tool. Each lend very well to forging by hand or under power hammer execution. Even the rivet hole punches are identical.

MISCELLANEOUS

Picture 303

Right to left: Handled hot chisel made of S7 tool steel. A 3-1/2 inch wide straight chisel made of oil hardened jack hammer bit. Top tool "V" cut of S7 tool steel. A wolf jaw tong made of 1 inch hexagonal jack hammer bit, 5-1/2 inches length per side of tong (11 inches total).

Picture 304

End view of top tools used to make the grooves of our wolf jaw tong (right of picture). The only missing top tool to make the entire tong is the rivet hole punch; pretty simple.

Picture 305 shown to the left.

The lower tool pictured is an oval eye saddle with handling bar. It is used to drive drifts through long eye hand hammers without damaging the eye exits. Many anvils, even up to 500 lbs do not have a hardy hole large enough to replace this tool. Here can be a good solution.

Picture 306 shown to the left.

At left is a tong which is used for weld scarf upsetting of round stock. Often as a scarf is made to fire weld a ring; it requires a slight upset to preserve enough material to complete a weld. As the stock scarf is hammered, it and the tong slips back through the smith's hand. This wide taper of the rein shoulder enables the smith to keep good control of all as the scarf is prepared. The cattle stomach magnet in the tong jaw is for size reference. A tong in this style can be forged to hold many different dimensions of material.

Picture 307 shown to the left.

The lock rings pictured are to hold reins of heavy tongs. They are cut off sections of track pin collars. They can be shaped to square, oval (as shown) or hexagonal to lock the reins of many types of tong. To elongate them, hammer the ring hole closer to closed. One inch sections of pipe in many different diameters work well for this purpose. If no pipe is available, a strap of flat bar stock or round stock is fire welded into a ring. The ring is drawn or shaped until it finds the ideal rein position when its intended tong is grasping the correct size of material. For us, the ideal ring position remains closest to the rein ends, as the greatest degree of rein flex is borne here to give return pressure against the ring. Reins which do not have much flex and return memory because of heft or type of steel from which they were made, will need their lock ring closer to the reins end (farthest from rivet).

When large billets (e.g. 3-4 inch round) are grasped in hand and rein ring locked, attention to time of grasp can become a factor. The tong jaws will pick up heat of contact from the billet and cause them to bend. The reins will slowly drift towards each other. Next, our billet is loose and cannot be held. Be cognizant of what the reins are doing - where the lock ring is when it was first pulled down to tighten!

Picture 305

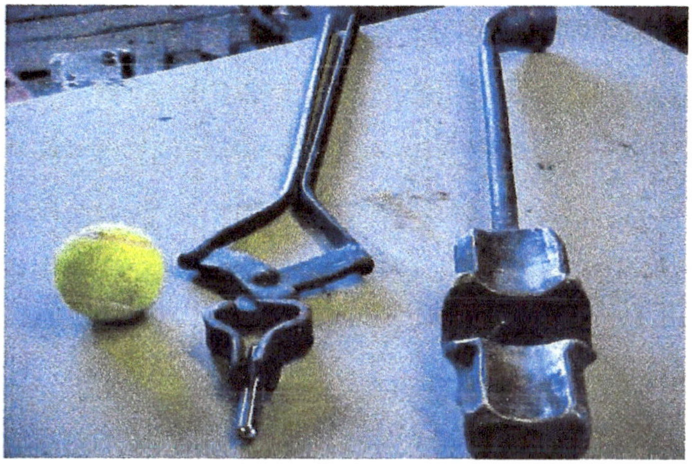

Picture 306

Picture 307

Picture 308

The tong pictured is for specialty use in building horseshoes. The hole in one jaw allows for the protrusion of a shoe with a heel calk (traction device). The opposing jaw has been cross hatched with a hot chisel for greater gripping abilities. These tongs can be built and sized to hold many different dimensions and projects such as rail road spikes, offset tennons, etc.

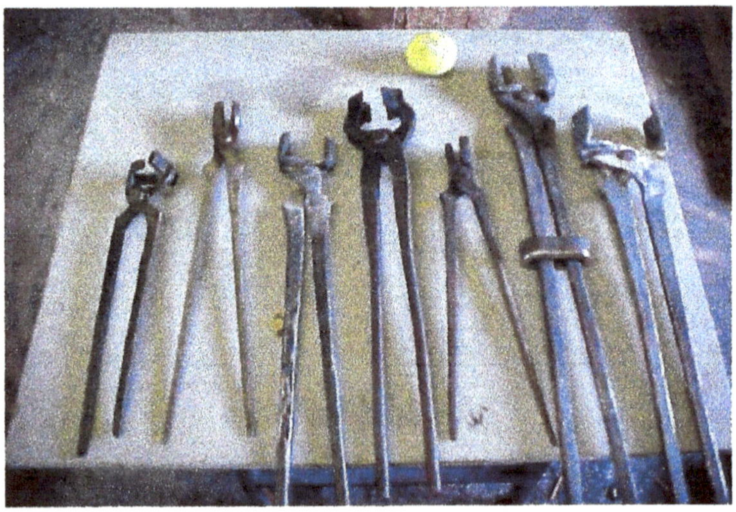

Picture 309

The two pairs of tongs to the right are light anvil makers. They are used in our shop to draw an anvil top heel from a 4 inch round billet. Next to the far right tong with lock ring in place is a tong to draw an anvil top horn after the heel is finished (including pritchel and hardy holes). Also for 4 inch round after all heel (or tail) dimensions are complete. A pair or two is made for all different anvil sizes.

Picture 310
Left to Right

-- Box jaw tong to hold 1/8 x 1 inch flat steel.
-- *"V"* jaw hollow bit tong with heavy shoulders. This tong weighs in at eight pounds. Used for 1-1/2" square, round, and hexagonal forgings. Tongs this heavy can be more difficult to manipulate.
-- A pair of shallow *"V"* bit tong used to hold hammer head forgings of 2" round. From jaw shoulders to the tip of lips, it measures only 1-1/4". Made of mild steel and kept very light for ease of manipulation.
-- Farriers flat jaw tong made of 4140 1" round. Will hold 3/8" flat stock. Each half of the tong required 5-1/2" to forge. From rivet to the end of reins, it measures 17".
-- Tong for redrawing a farrier's fuller (shoe creaser). The median tip fits into the fuller eye for security during the redrawing of its creaser edge.

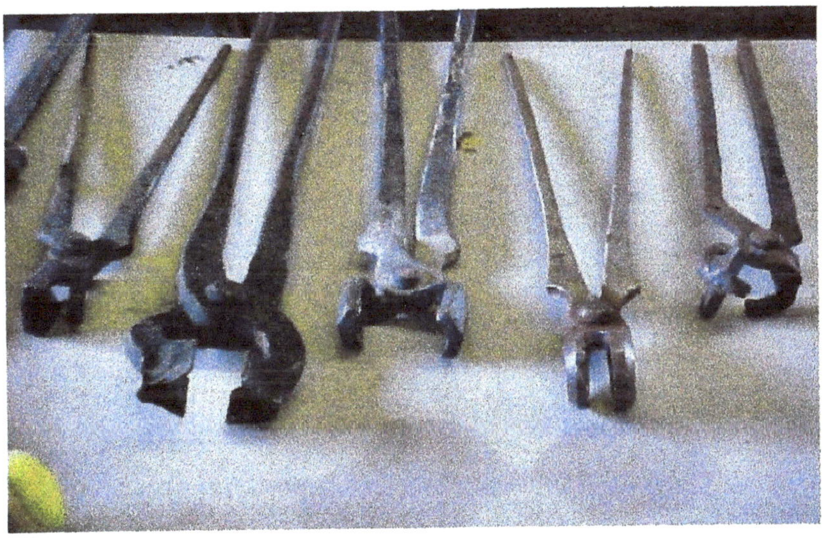

ANVIL REPAIR

Pictures311, 312, and 313 shown opposite page.

Shown is one of the old forged farrier anvils. It came to the shop for repairs. Missing two thirds of its faceplate and the remainder still firmly secured, a plan of attack had to be formulated. The pieced together approach was decided upon. Three quarter inch round H-13 was cut into 5 inch lengths and forged to flats of 3/8 inch thick. They were left slightly thicker than the remaining faceplate to allow for finish grinding. Pritchel holes of the exacting sizes were hot punched where needed. The square hardy hole was surrounded by billet ends and sides. Gaps between billets are left for secure arc welding to the anvil top on all billet sides. Portions not welded are inside and at the exposed upper crest of any holes. If internal hole welding is ensued, clean up filing then becomes excessive and provides negligible additional security, regardless of anvil size for this type of repair. DC arc is used here, 2110 build up rod, and final passes and edges are made with 1105. These are high tensile rods not hard-facing. After each pass, the laid down bead is hand sledge hammered to work harden and remove scale. For the pieced together method, all billets are forged, air hardened, placed in their final destination and tack welded in position before heavy passes of welding are made. Each piece is clamped in position for welding to prevent loose spatter from creeping under the plate billets. The heat from welding tempers our plate billets here. All is allowed to air cool when finished.

No two anvils brought in for resurrection have had an identical procedure to rebuild. Good thought and study should be involved before any commencement begins. Consider the brand of anvil (cast or forged)? Where the damage is? Underlying stress fractures? Was any previous repair or defect evident? Wire brushing all surfaces with a cup brush in an angle grinder may reveal some skeletons. Consider if the repair is worth making or should all be left alone?

Picture 311

Picture 312

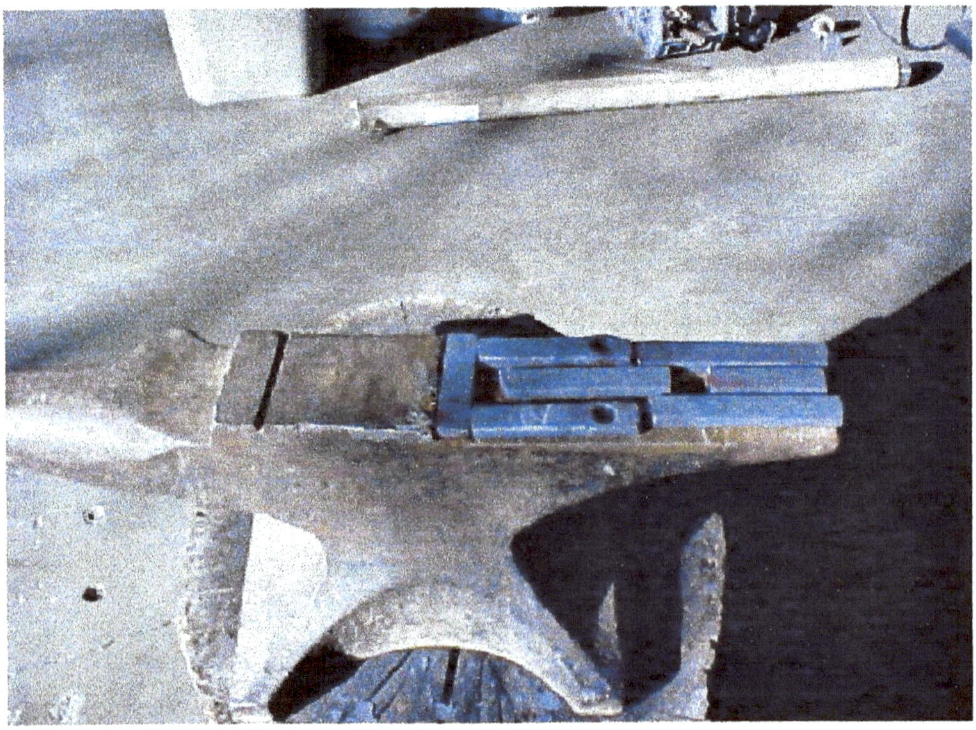

Picture 313

Picture 314

Clamping of all forged billets so they may be tack welded before heavy passes are attempted. This prevents any unnecessary movement and therefore misalignment.

Picture 315

Wooden blocks are used for anvil positioning during repair work. Many arc welding electrodes are gravity specific (will flow where gravity dictates), and require the base metal/steel to remain supportive. By positioning the anvil accordingly, the welding bead will lay down where it was intended to be. The anvil shown weighed in at 158 lbs for size reference.

Picture 316

Picture 317

Picture 316 – Shown on opposite page.
Yes, this is the same anvil! Welding completed, work hardened with an 8 lb sledge before cool.

Picture 317 – Shown on opposite page.
Finish grinding is commenced and completed only after the anvil has air cooled and cold. Some repairs will need to cool slowly in wood ashes, lime or vermiculite. Each works equally well though ashes may be most cost effective if a wood stove is used for household or shop heating. The anvil is ground in small increments of each section so as not to overheat any one portion. Stress fracturing of welds and/or anvil can occur when one segment shows the color of heat due to grinder friction. Blue/purple, straw/brown and any colors above, below, or in between will tell that the grinder has overheated the steel.

Our grinding process is progressive in selection of grit. The start is usually 36 grit. We proceed to 50s and 60s of a flapper disc type in an angle grinder. These are all hand-held. We wear safety glasses, ear plugs, and a breathing mask. Ventilation should be good and no dry grass is nearby to start a fire from the sparks. The next grit is an 80, and lastly a worn out 120. Finer grits can also be used to finish if an even more brilliant mirror polish is desired.

For flat surfaces, a grid pattern is used to end in a level product. Cover the entire surface as is

//// then \\\\ next |||| and lastly ═════

written with the angle grinder, as many passes as are needed with each pattern. Much care is taken not to overheat any one section.

Picture 318

Pictured is a security bracket forged before the turn of the 20th century. It was removed from a 225 lb anvil known to be made between 1850 and 1860. The anvil was in for repairs of a crack which ran vertically from the face and step toward its waist. This bracket was all that kept the horn from falling to the floor. It straddled the cutting table and traversed the anvils near and far sides. The nuts and strap were secured under this anvil's tail and waist. Ingenuity of the old school.

Picture 319

A selection of anvils kept in the shop for repairs. When working in your shop or others, it may be of great benefit to try using many different types and styles. Try and consciously find aspects of an anvil that appeal to you. Is it the sharp or rounded edges that appeal to your general work type? Could a wider or narrower face be of benefit? Is the horn long and pointy or too short and stubby? The short and blunt horn can make getting into tight turns harder for some smiths. The chief consideration for many is simply the weight of the anvil. But how is this precious tool proportioned?

Picture 320

Many of our repairs to anvils are to the edges. Regardless of shape, size or weight, that has been a common denominator. The anvil should be considered an important and indispensable tool rather than a low slung wrecking ball. Most of the damage observed is not from manufacturing defects, but user error. Try to rate and measure all hammer blows to prevent undue abuse delivered to your anvil. If larger projects are to be tackled regularly, it will serve to give the 100 lb anvil a rest. Try and find that larger 200-300 lb you always wanted. Not only will the larger one recoil your hand hammer better, but the savings in time and orthopedic medical bills for an aching elbow are more than proportionate to justify the anvil addition.

Picture 321
The anvil pictured must have weighed over 200 lbs when new. The brand reads "Trenton" and had no bottom from the waist weld down. This proves that even a throw away to some is a vision to others.... The base, though now from a transfer gear casing, is back in place.

Large double horned anvils can be hard to find in any condition. Why don't we make one? They allow for easier leveling of material in narrow spaces. Below is an English anvil conversion of material from traditional to the old double-bick type. A longer and larger round horn on one end, a longer tapered heel is added to the other. Good DC arc welding skills will be needed here. Chip all flux, scale and wire brush between electrode passes. The addition pieces are H-13 tool steel, air hardened then immediately arc welded into place with 1105 rod. The heat of welding tempers these pieces.
Extra Rod is layed down to allow for a clean finish grind. Work hardening of each heavy pass is done with a 8 lb sledge as soon as it changes from red to black color of heat. All is allowed to air cool.

Picture 322

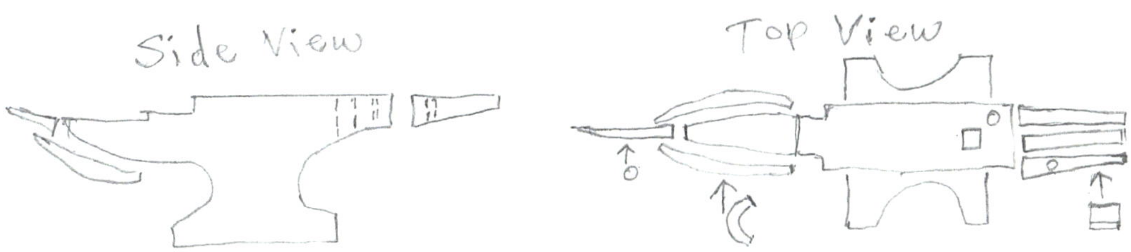

Parts and pieces are forged for what will become a large (300 lb. plus) double bick anvil during the renovation process.

Anvil Additions (traced actual size)

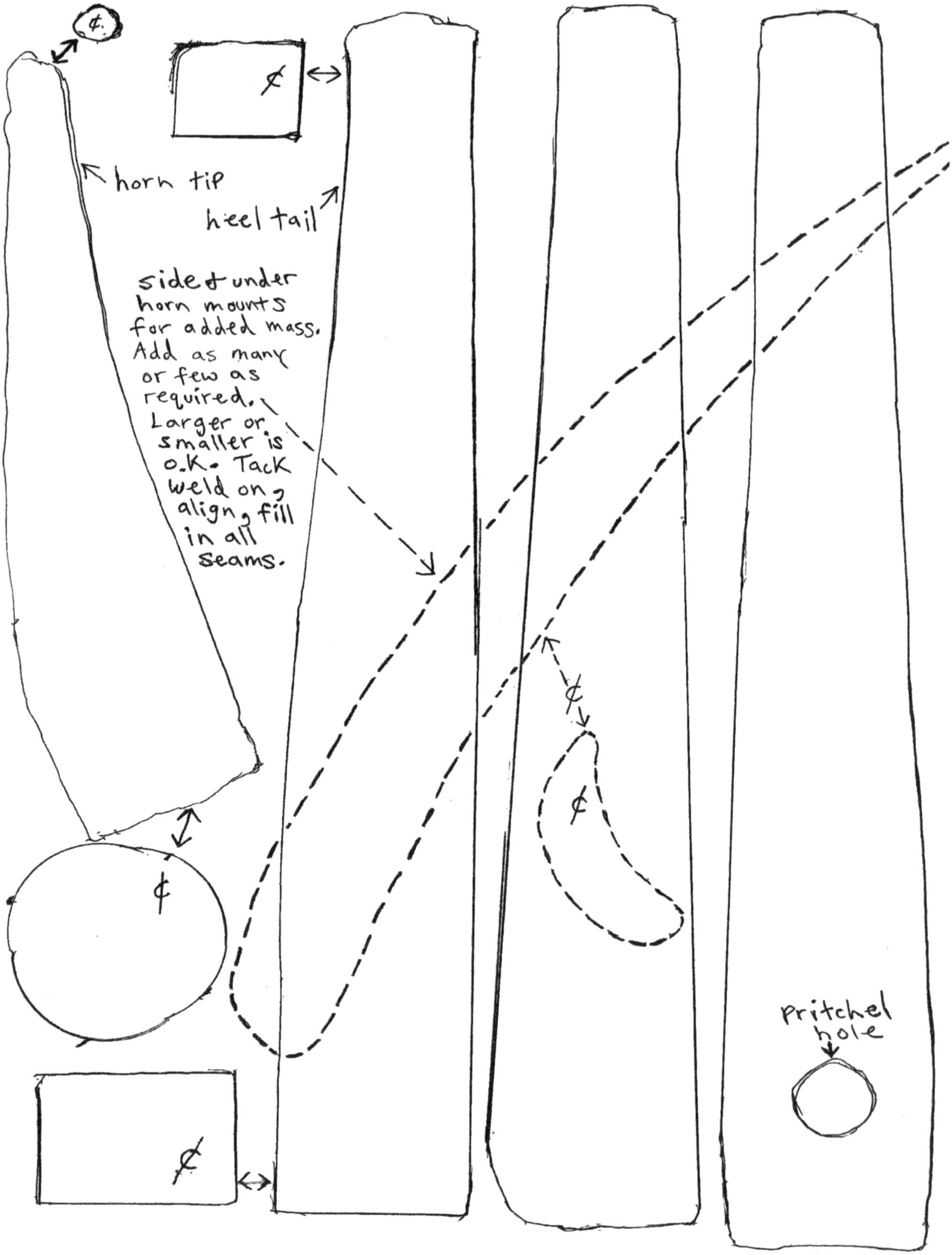

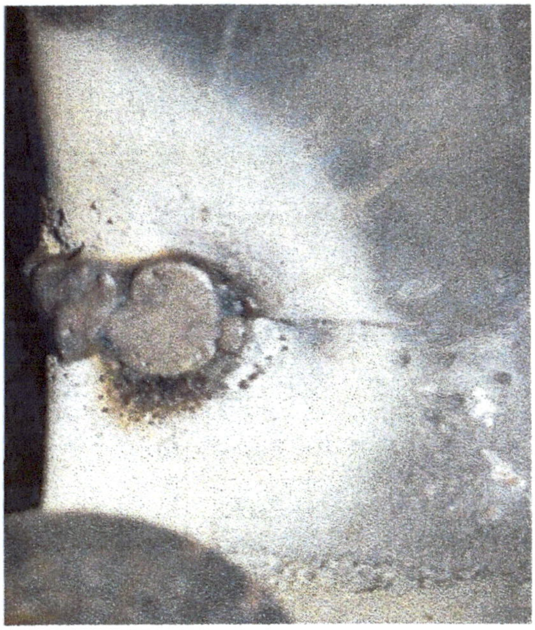

Picture 323

Picture 324

While making a repair on an anvil through arc welding, the heat produced may show other suspect regions. This shows a waist weld under our anvil horn separating from the heat generated during a face plate repair. Stop right there. Let the anvil cool off a little. Weld up the separating portions. A heat differential is enough to push pieces apart. Even if all parts are at the same heat, they have expanded with the heating. Different thicknesses expand at different rates as well.

There is even hope for some of the most abused anvils. This english anvil had very deep torch cuts through its face plate and heel. One cut was so deep, it was decided to drill through completely and have two pritchel holes result. This sequence is of the same anvil - start to finish and made ready to re-enter the work force.

Picture 325

Picture 326

Picture 327

Pictures 328, 329,
330, 331

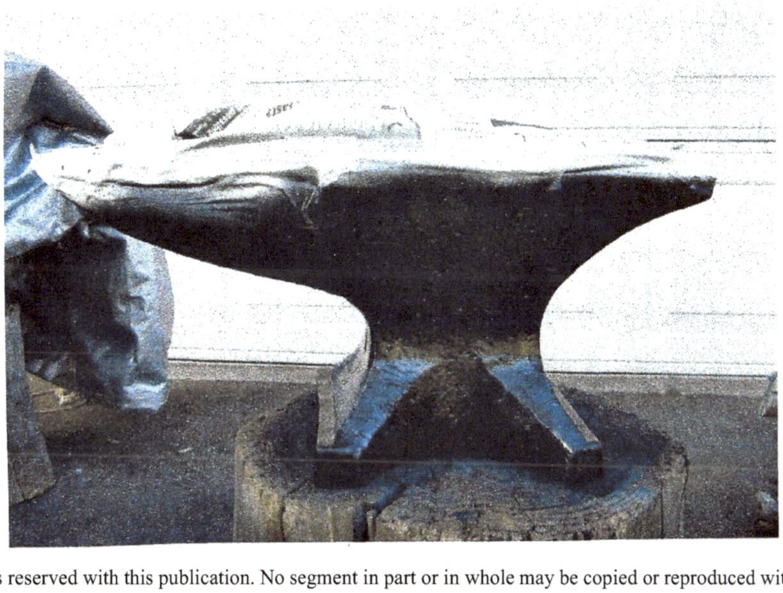

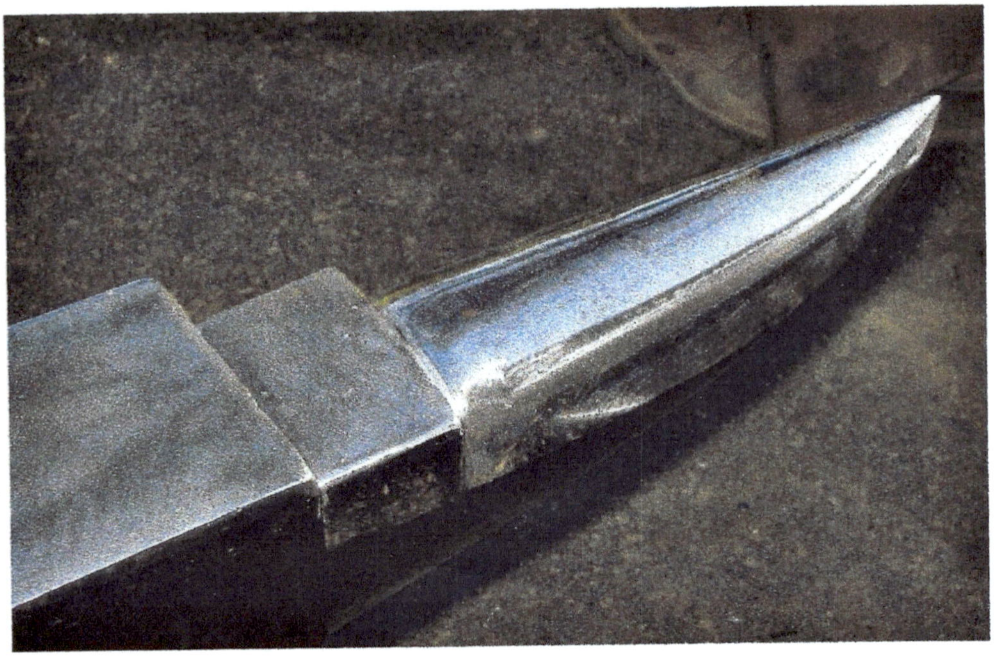

Picture 332
The finished anvil ready for service.

Picture 333

DIFFERENTIAL QUENCHING

DIFFERENTIAL QUENCHING

On many an occasion, a mild steel remnant has been handy for tong blank length, but not wide enough to provide a sturdy rivet hole boss.

The steel is lightly marked with a cold chisel to depict the location which will be upset to provide additional width. If thickness is also desired, both tong blank halves should be marked to accommodate this, and hence they will become a pair.

Picture 334
Both halves are marked as shown and placed in our forge for heating.

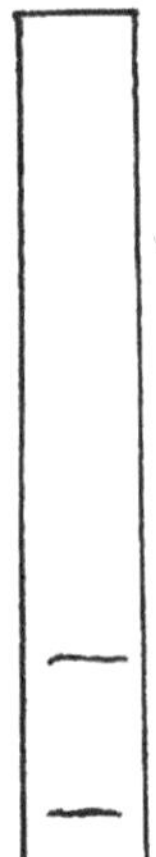

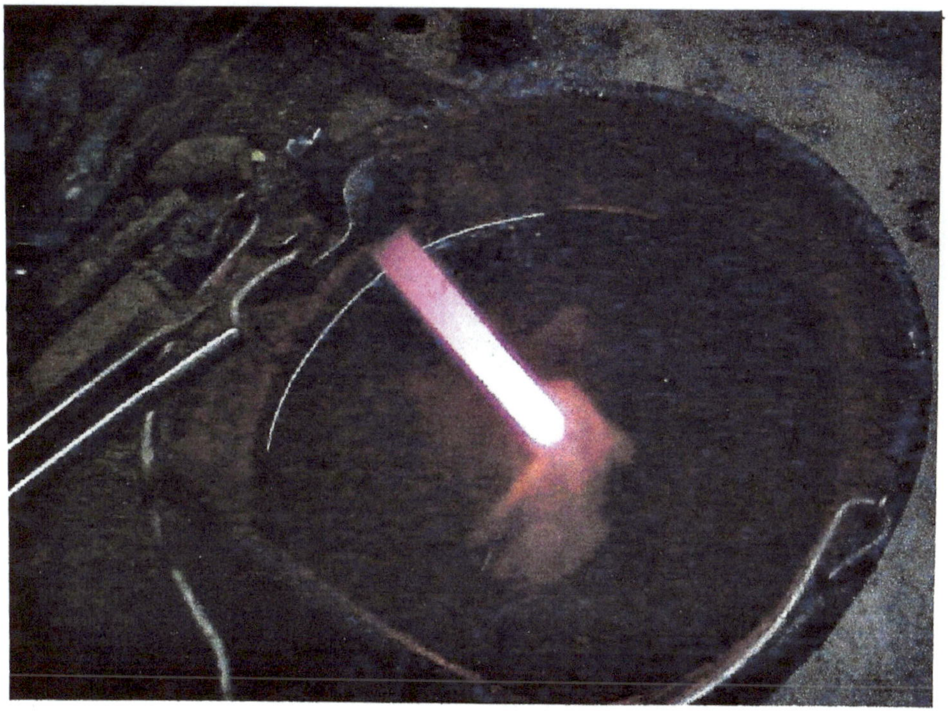

Picture 335
At an orange heat the billets are drawn from our forge and quenched in water up to the distal cold chisel mark, farthest from our body.

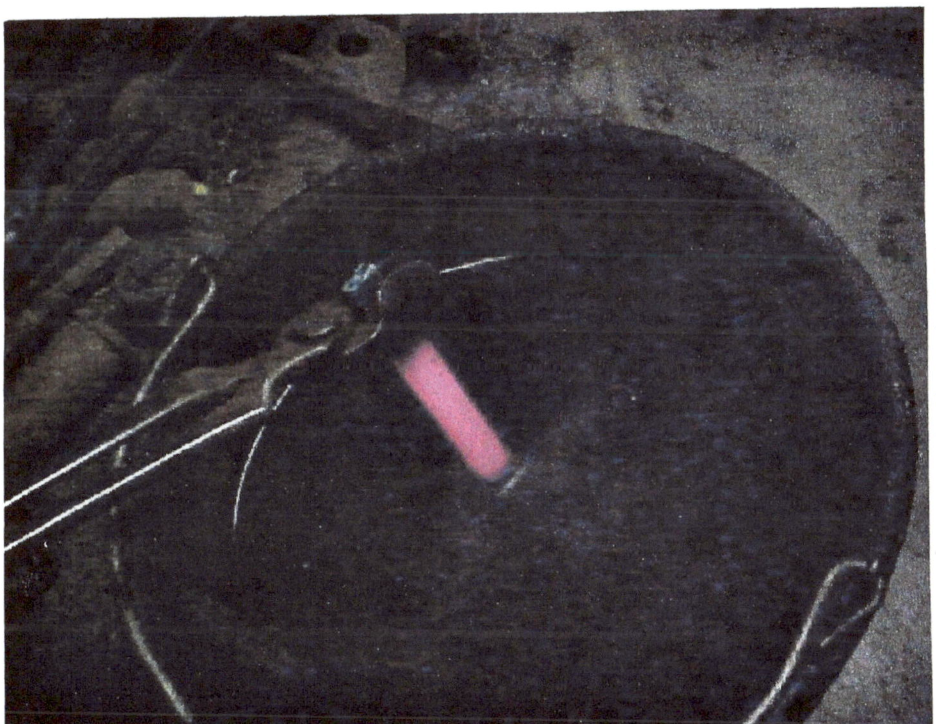

Picture 336
Quickly the billet is then turned around and quenched up to the remaining cold chisel mark.

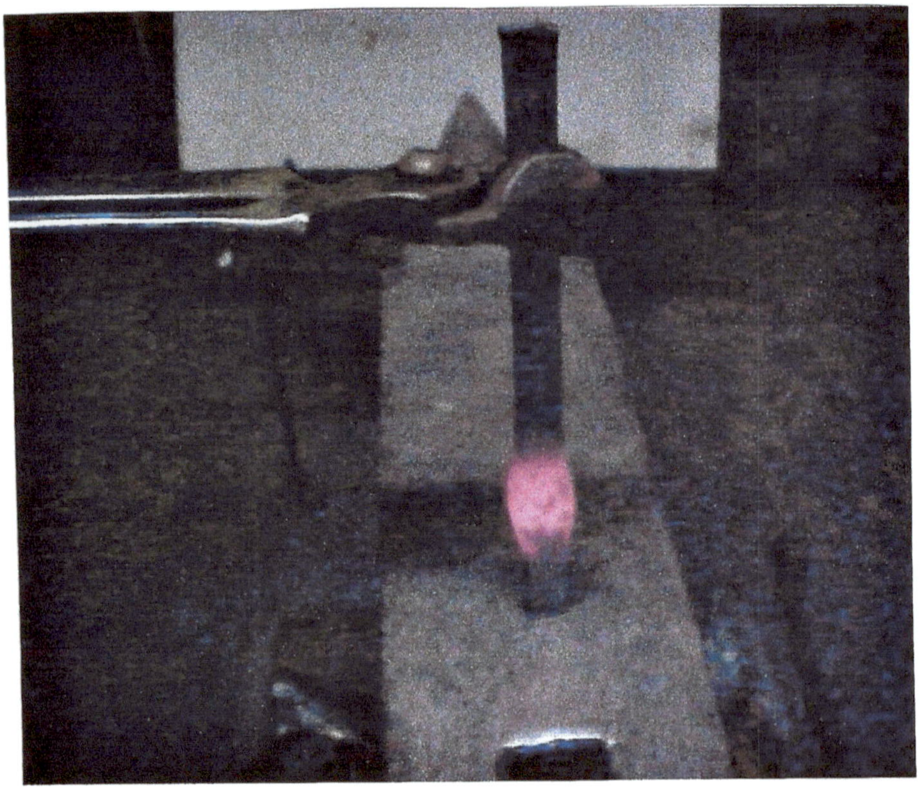

Picture 337
All that remains at a forging heat is a segment between our cold chisel marks. Now the hammering begins as shown. When any bends or bows occur they are immediately corrected.
The progression is a few blows endwise and a few blows flatwise, very little edgewise, only to create straightness to continue upsetting the forging.
The end result will look as shown. It's O.K. to take more that one heat to complete this satisfactorily.

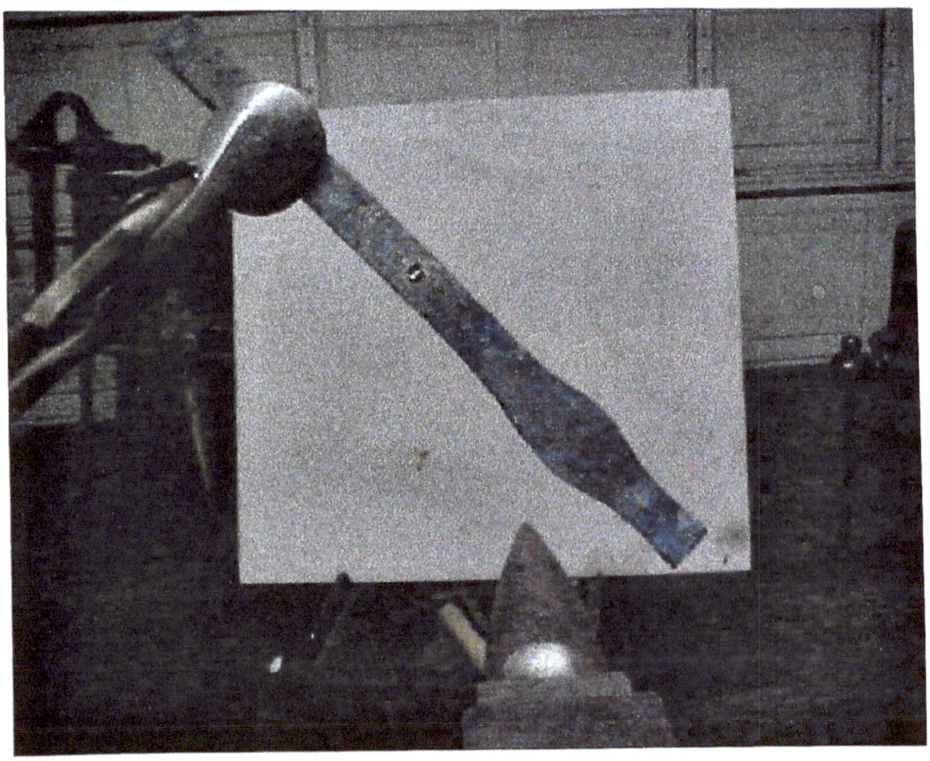

Picture 338
The upset bulge is developing here as was intended.

Picture 339
Here is another means of correcting straightness during forging over the hardy hole.

Picture 340
Shorter pieces are also upset in this manner. Note that the hot segment is usually kept closest to the anvil beneath. This piece was upset to add mass for a jump weld to tong reins.

DOUBLING OVER METHODS

DOUBLING OVER METHODS

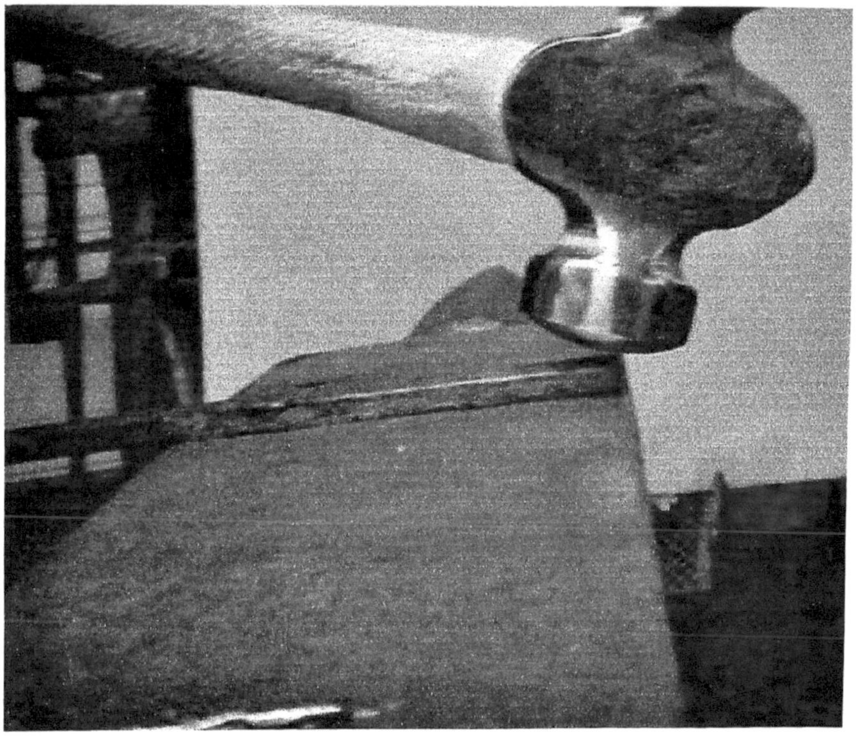

Picture 341
Preparing a tong billet for doubling over to add jaw and rivet boss mass. The hammer blow is far edge half face rounding. This gives the smith a good start at the scarf before any folding and welding is commenced.
As the billet gains in width it is turned up on edge and hammered with the rounding, then a flat face back to uniform width.

Picture 342
The scarf finishes out as shown.

Picture 343

When the stock at hand has not enough mass for an ample tong jaw or rivet boss, this is how I do it. The mild steel end is scarfed as shown. The piece is hot cut two-thirds of the way through for ease of folding. After doubling over, we are ready to, sprinkle with borax flux at a cherry heat and fire weld all-together. If the billet material is made of carbon steel (coil spring, axle, sucker-rod) a slim piece of mild steel should be slipped into the seam area at the time of folding.

This will enable a better fire weld. Carbon steel does not fire weld perfectly to itself. Even with practice we can get it done, but not without consequences. The higher the heat required to enable its semi-molten welding state may only render fracturing/ molecular destruction. Mild steel in between will help though.

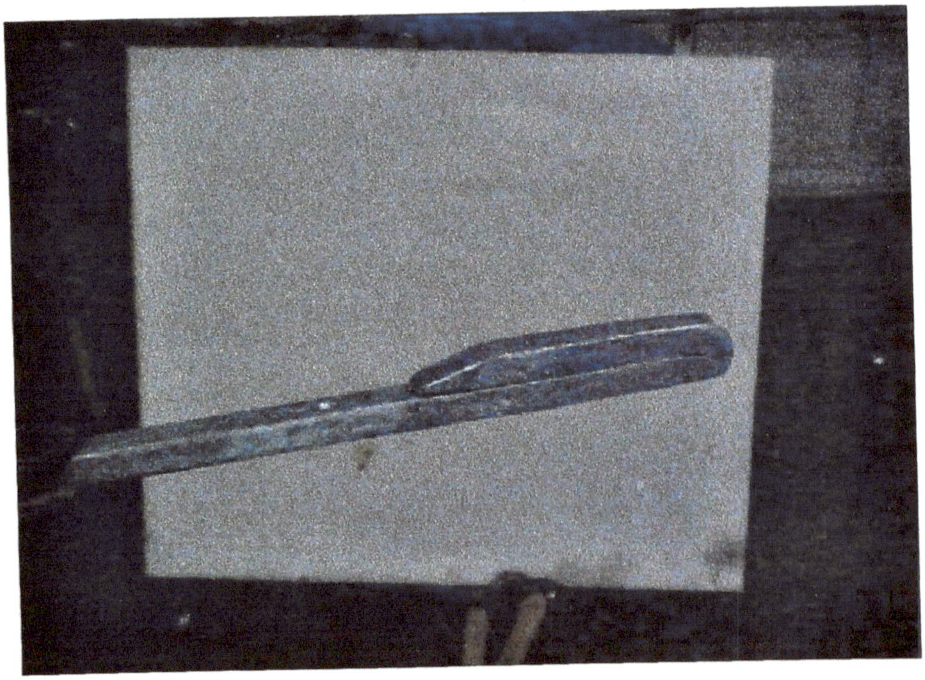

Picture 344

MILD STEEL vs. CARBON STEEL IDENTIFICATION

MILD STEEL vs. CARBON STEEL IDENTIFICATION

If there is some uncertainty as to which grade or steel composition is on the anvil, here is a simple test to get us in the ball park. Present the cold steel to a bench grinder or belt sander. A higher carbon steel will show fine feathers trailing the sparks. We will at least know now that we don't have mild steel.

Picture 345

Carbon steel grind test

For mild steel identification the same process is repeated. Present the material cold to the grinder for spark observation. The material leaving the billet will appear coarser and heavier in granule composition. Spark brightness can be no less evident; we are looking at the integral definition of dispelled projectiles.

Picture 346

Mild steel grind test

Safety glasses should be worn at all times for these tests. Dark welding goggles may also be needed should the billet be titanium, which will show bright white sparks.

JUMP WELDING

JUMP WELDING

There are innumerable means and methods of achieving a fire welded union between two separate steel pieces or billets. Even within a single shop, what is routine protocol for one smith may have the next stand on their head for even close to satisfactory results. Most successes are a product of adequate preparation of material before their union. Clean surfaces as they are brought together. Proper attention to union mass to gauge heat tolerances is paramount. Trusting ours eyes to decide when enough heat has been imparted to the billet, trusting our hands to get the billet from fire to anvil or trip hammer quickly and efficiently to begin to strike the first tell tale blow.

Jump welding by definition can have two different meanings in our shop. The pieces are heated simultaneously but separate from one another and later brought together on the anvil for hammering. They can also be forged to create a means of union before any weld takes place. Which ever preference exists, this is how we separate the ideology. This is one of those segments where form can follow function. If the piece/pieces are too heavy, intricate or awkward to fit in the forge unionized; they should be easier to join after extrication from the fire. A helper to steady one piece while the smith holds the other and swings a hand hammer is often the sequence here. If the first piece will hold still on the anvil the smith can proceed alone.

Because I work alone at many of the shop projects, some innovative means have come to evolve to achieve a dependable jump welded union. Challenging ourselves to find the appropriate union method for the project at hand will help grow roots for this work.

These are but a few means which are used regularly for union before a welding heat is imparted. One was developed, the others are from the past revisited; can you tell which is which? Each has become a tong blank.

WHICH STOCK MADE WHAT?

Billets and tong blanks can be paired up based upon steel availability in the shop on a given day.

Identical stock next to tong blanks from which they were forged.

Picture 347

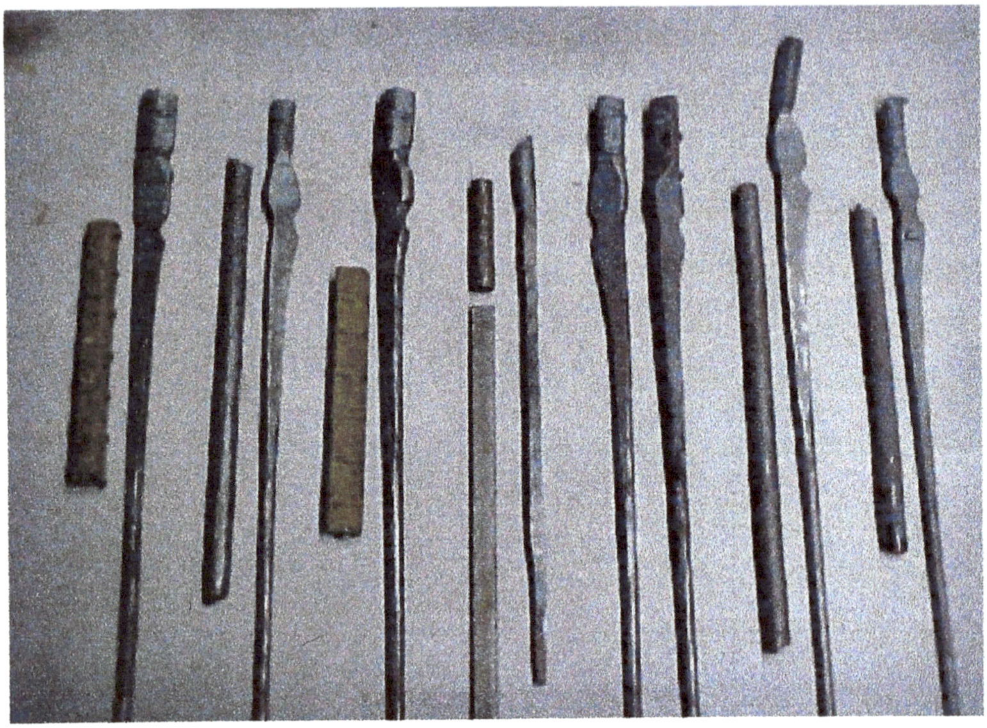

A **B** **C** **D** **E** **F** **G**

A- 3/4 x 6" rebar
B- 5/8 x 9"coil spring straightened
C- 1 x 6" rebar
D- 5/8 x 2 ½ " coil spring jumped onto 5/16 x 3/4 x 9" mild steel
E- 3/4 x 7" rebar
F- 5/8 x 10" coil spring straightened
G- 5/8 x 8" coil spring straightened

SCARFING TO WELD

Picture 348
Our steel is heated and held at the near side anvil edge, the hammer blow is half face flat. First setting the anvil edge to form a shoulder.

Picture 349
The steel is raised as our scarf forging progresses gradually. It finishes parallel to the anvil face. This piece will become a tong rein.

Picture 350
Here is the blow with hammer in the correct place at time of impact. A half face flat near edge blow is employed.

Picture 351
During the scarf set down, unwanted width may develop. This should be addressed as it occurs. Attempting to gather up all the excess width after welding may cause undue stress to this section. After welding skills develop, the smith can then judge mass requirements and make forging adjustments.

Picture 352
Our scarf is brought to the anvil full face and hammer full face and forged back to uniform width. Near edge half face blows for clean up are again resumed to restore stock uniformity and blend any imperfections.

Picture 353
Here is a side view of the progrcss so far.

THE DOUBLE FISH HOOK

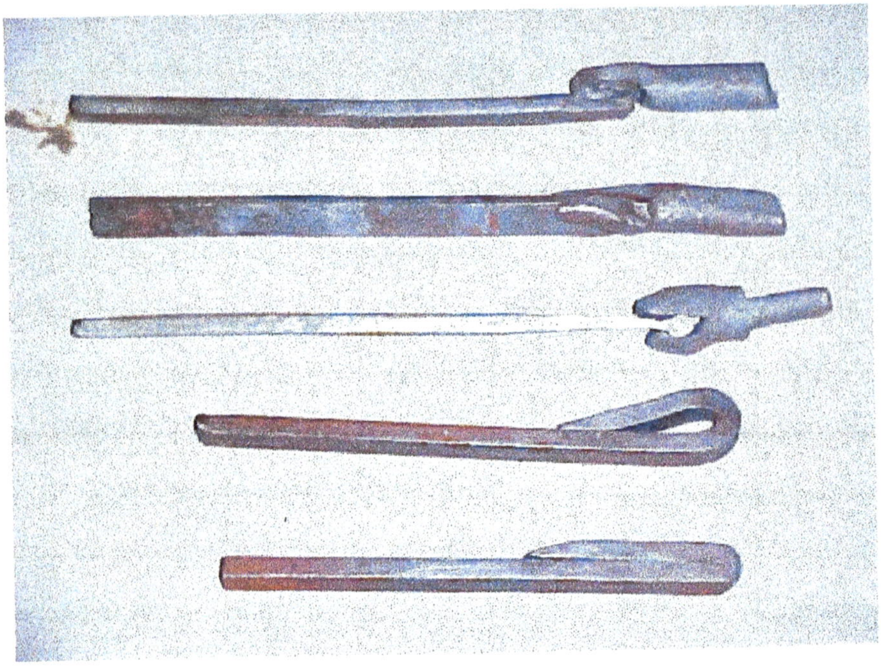

Picture 354
Double fish hook

Split

Cleft

Curl scarfed

Full lap

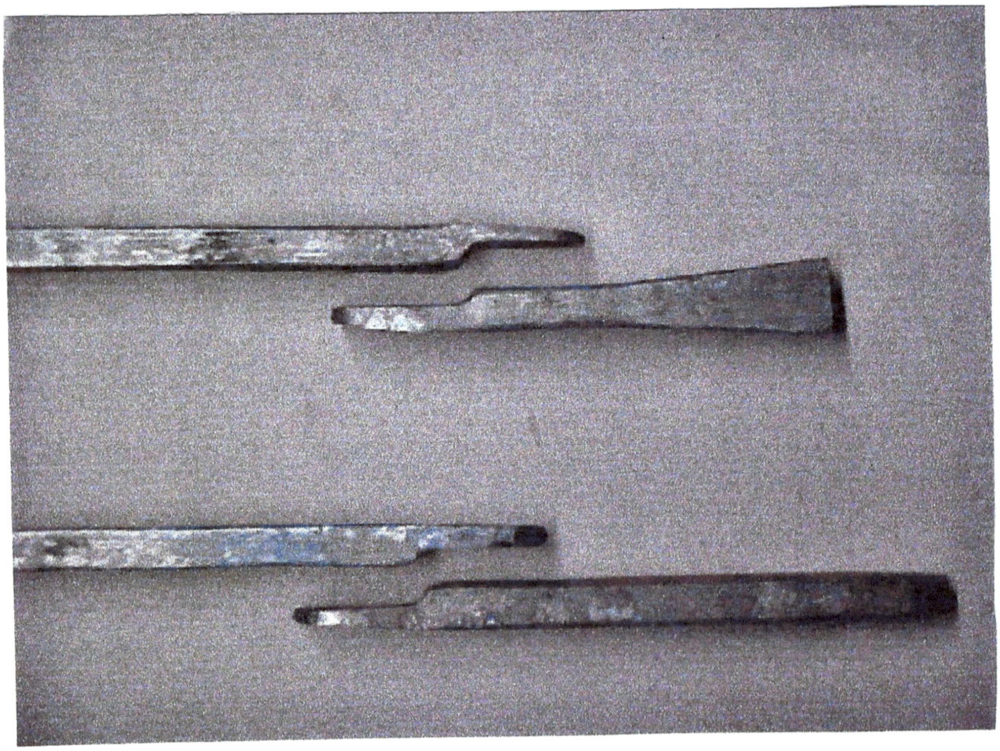

Picture 355
This is the double fish hook set up- it is started much as a shouldered lap weld but each is turned to hold its mate tight before welding.

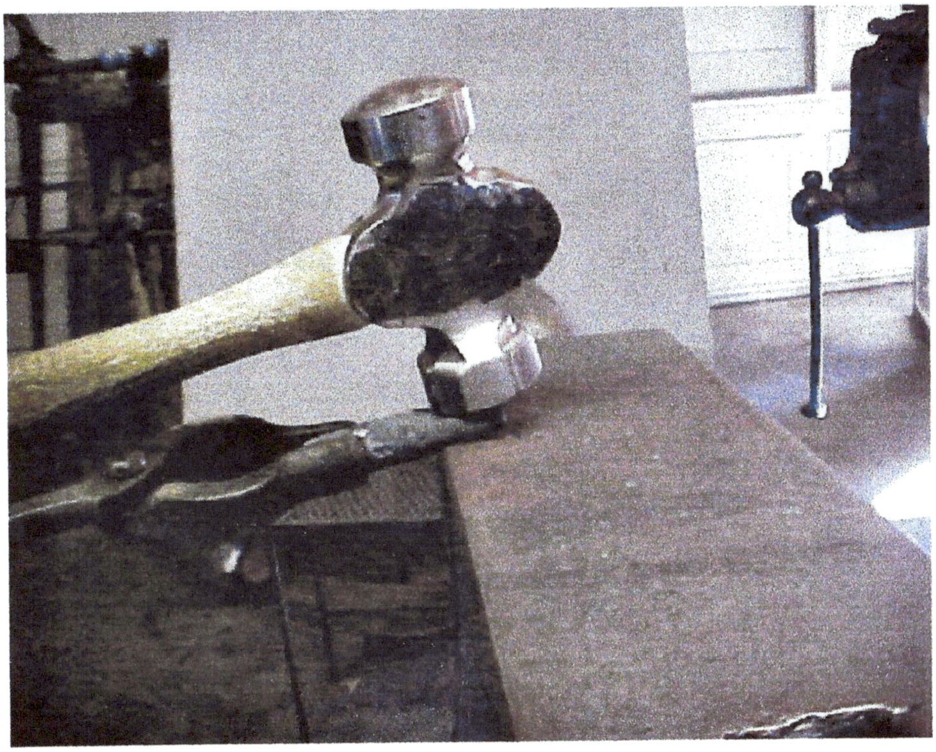

Picture 356
The tong shoulder to jaw lip piece is then cut and also scarfed identical to its rein mate. Here the piece is 3/4" round, 3-1/2 " of coil spring. Because of the spring's carbon content and tensile strength it refuses to move even after a yellow heat. We use a heel blow to pinch and start the forging, then proceed as normal.

Picture 357
The finished fish hook is then turned on our horn.

Picture 358
The two pieces are now brought together to check for curve symmetry and void depth suitability.

Picture 359
When all looks well, the pieces are entangled at the light cherry heat and set down with a full face blow.

Picture 360
The double fish hook will look like this with a minimum of air space within. We can now give it a light heat, wire brush, apply flux and take it back to the fire for a welding heat.

This means of welding requires a little more forging prep prior to their union but has many advantages for the smith. One can do it by themselves. Mechanical advantage does occur in addition to the weld. There is less need to run from the fire to the anvil in hopes that enough heat will remain to stick one piece to the other. If the smith should miss the welding heat (too high or too low) the pieces can be cleaned up (brushed) or in worst case scenario, taken apart and tried again without total loss of material.

THE CHECKED SCARF JUMP WELD

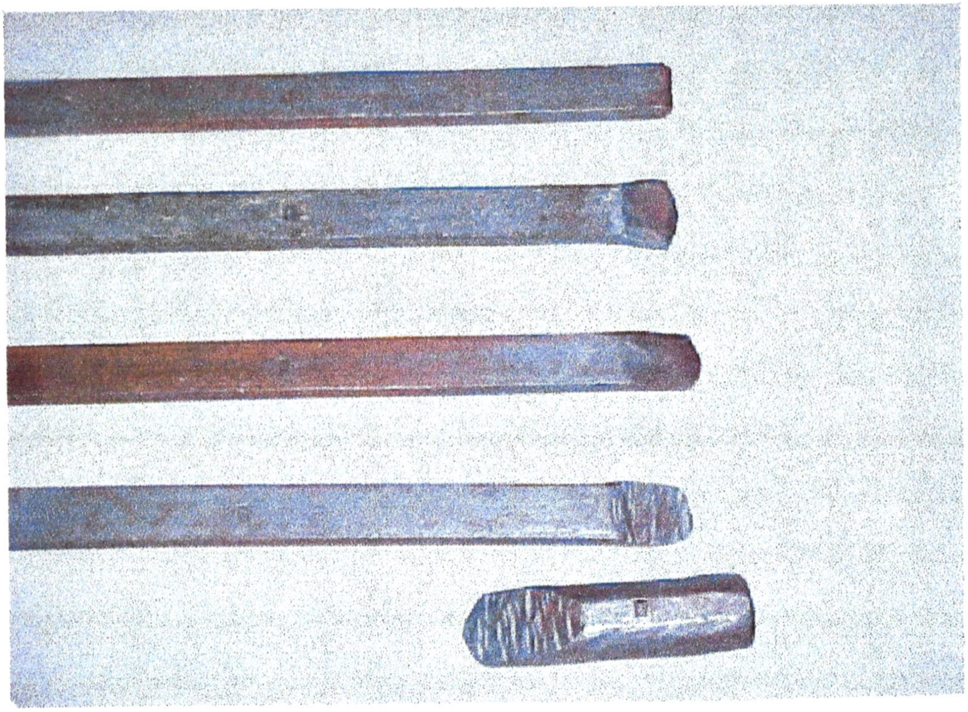

Picture 361
The progression here is top to bottom, where 5/16 x 3/4 inch flat stock is made ready to jump onto 1 inch hexagonal.

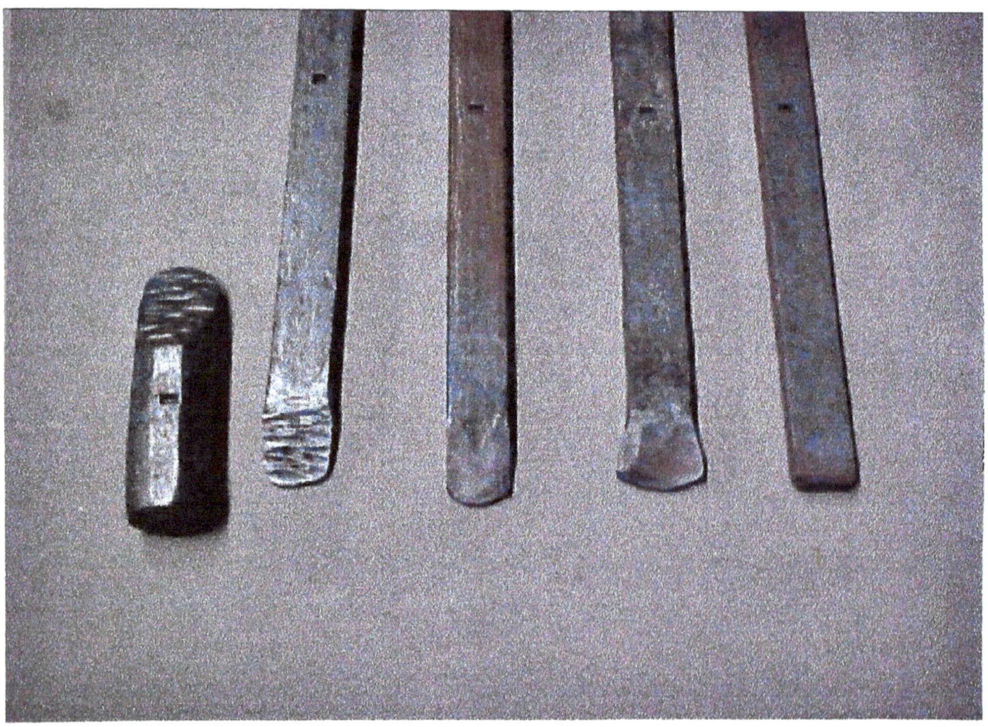

Picture 362
The progression is right to left with steel end prep of importance. The entire forging sequence of flat stock was accomplished in one heat.

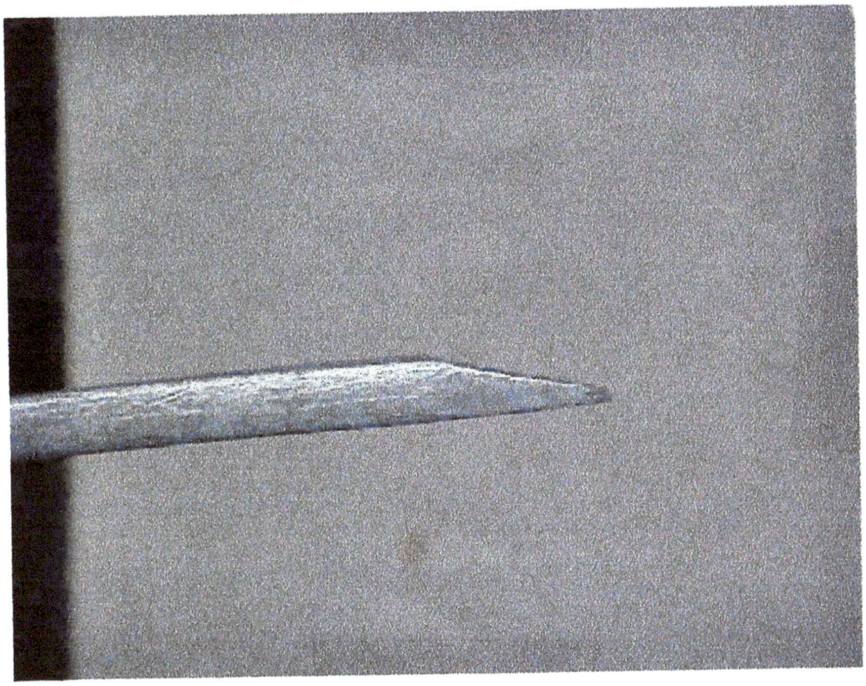

Picture 363
Our side profile will look as shown for the flat stock. It is now ready to have its checks put in.

Picture 364
The far edge heel blow is all that is needed. Used here is a sharp edged flat side of the rounding hammer. A 1/4 inch edge radius hammer will work here as well.

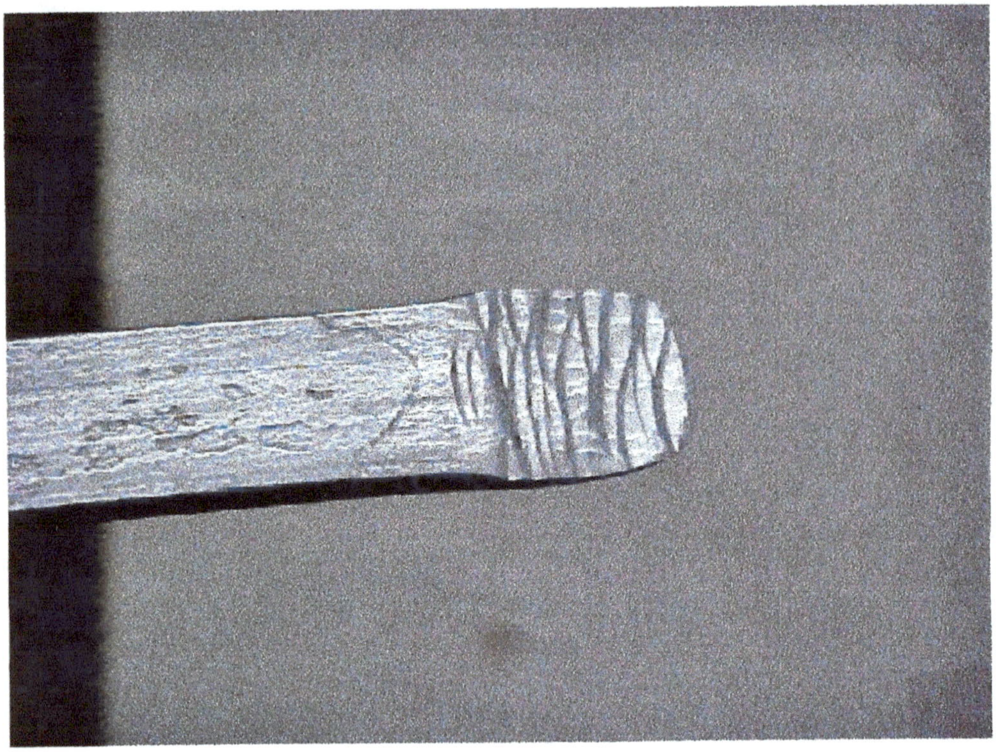

Picture 365
A close up of the checked scarfed surface. Why check the surface? There is less chance of the union slipping apart as we are hammer and fire welding it together with its eventual mate.

Picture 366
Here is the start of a one inch round three inch long addition. It will be jumped onto by the above flat stock. We are creating a <u>flat</u> midline so that it <u>will</u> <u>not</u> roll during welding.

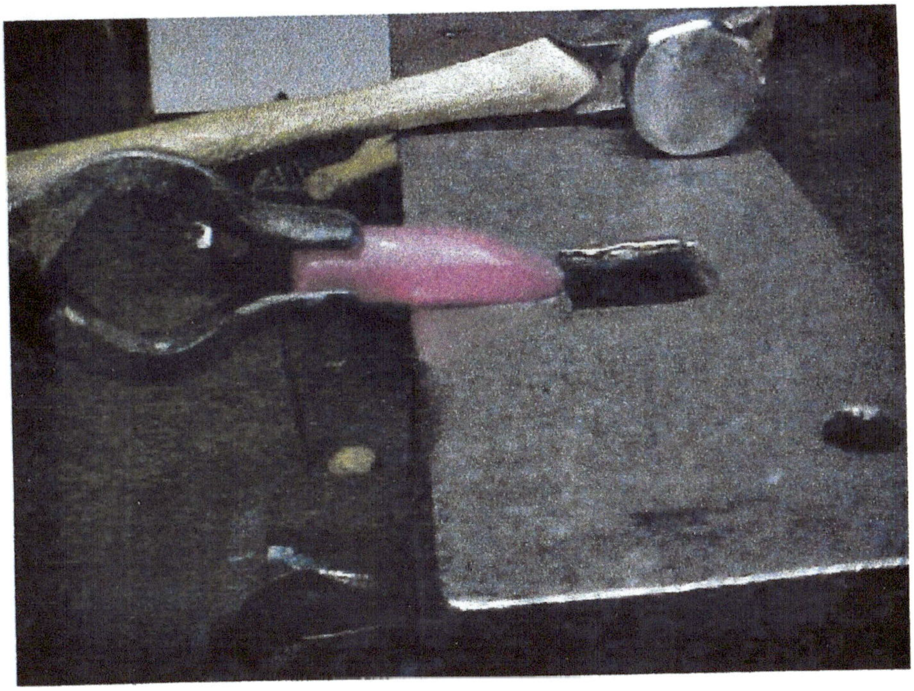

Picture 367
The round stock is now ready to receive its scarf and checks. With short pieces it helps to make use of the hardy hole to reach the distal steel tip (farthest from the smith). This will save your working tong from abuse and provide efficiency to the forging.

Picture 368
Both pieces are brought from the fire simultane-ously. The larger round stock section is set first.

*Note the back-up brace hardy hole mounted.

THE SPLIT WELD UNION

Picture 369
There are two thought processes to consider in preparation for this weld union. Do we drive the steel through the chisel hardy to create our split? (Brass hammer used here)

OR

Do we drive the hot chisel through the steel?

All dependent on the steel dimensions….

Picture 370

Picture 371
When pursuing the split weld cut using a top tool some care precautions should be made. Here a copper plate is placed under the work. As the final cut is made we do not damage our hot chisel or anvil.

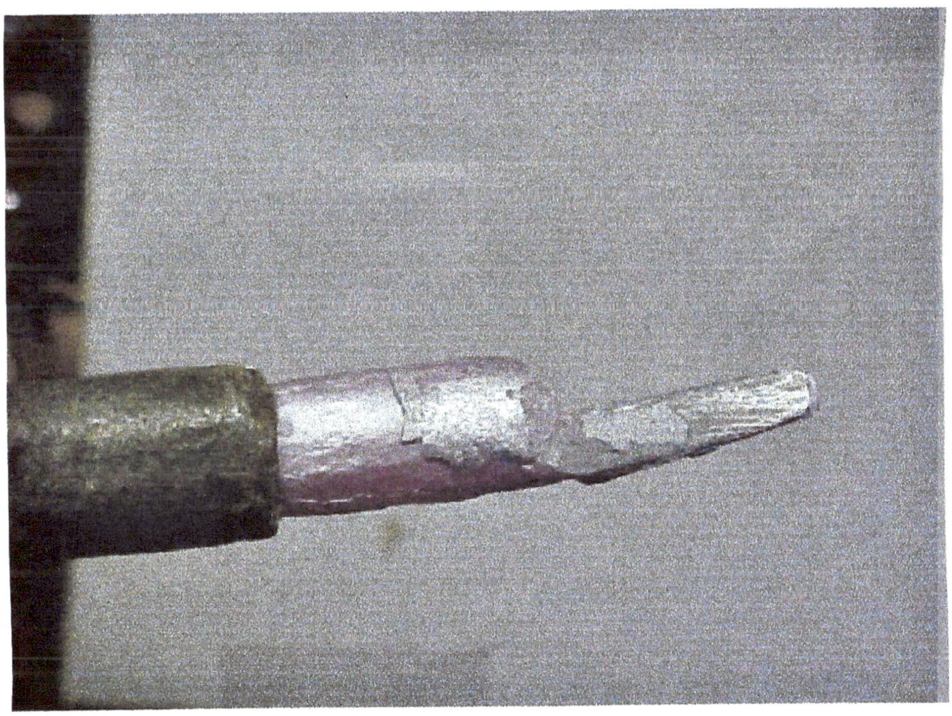

Picture 372
A side view of this forging before and after the split is cut clean through should look as shown.

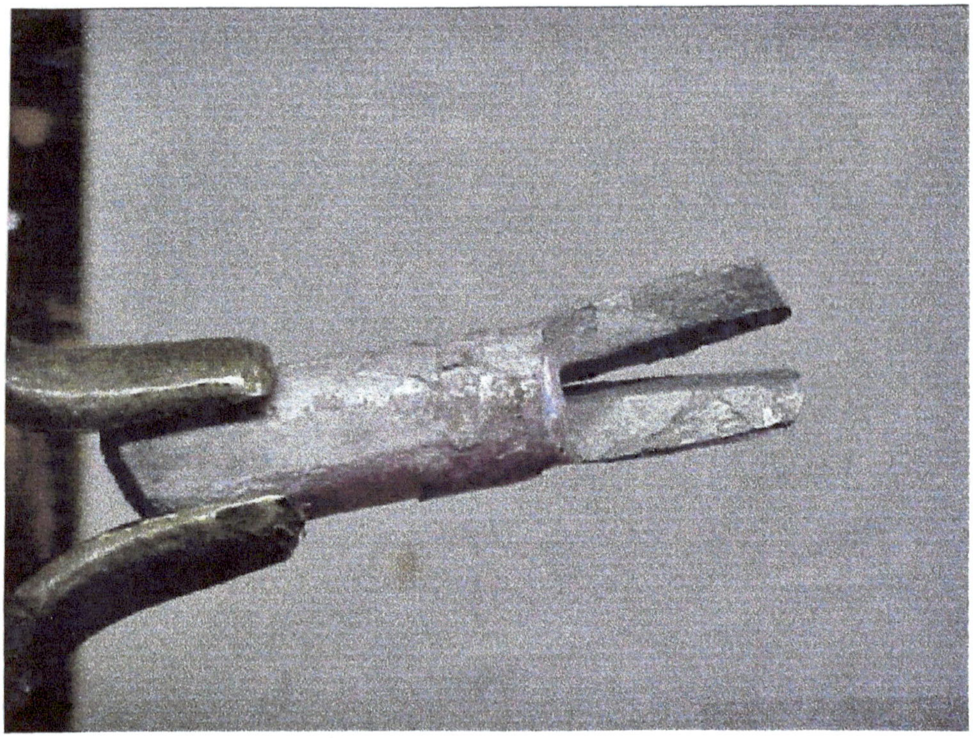

Picture 373
Top view of a clean split weld prep-cut. We are now ready to take another heat and part our ends. This is not as a scissors moves, the tabs are prepped and moved to and fro rather than sideways.

Picture 374
The ends are worked over our anvils near and far edges to separate and prepare each for the best welding contact with its mate.

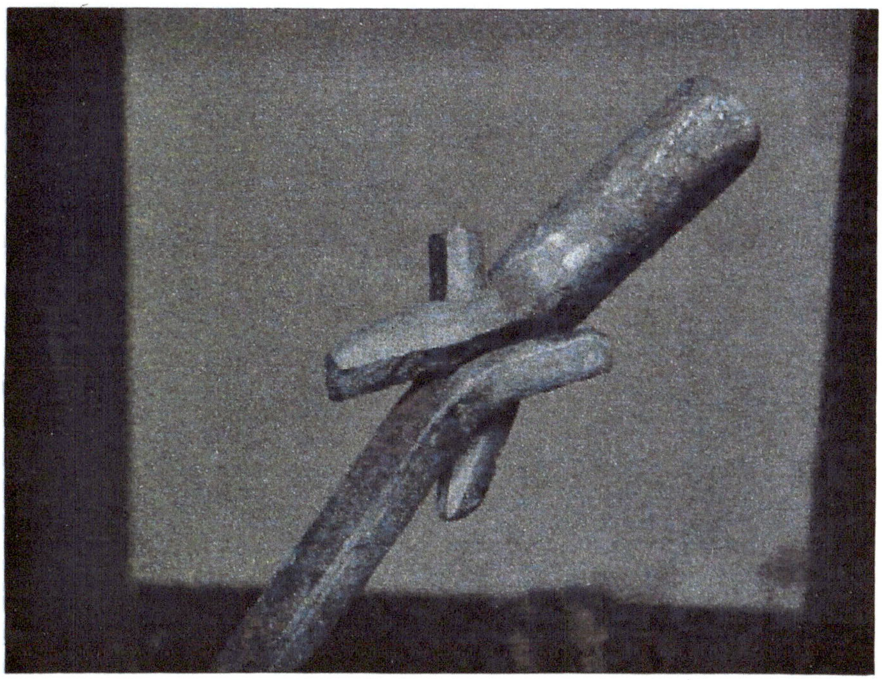

Picture 375
Our pieces are now ready for a union check. When they butt up to one another comfortably we are ready to take a cherry heat and forge the split ends down lightly on the anvil face.

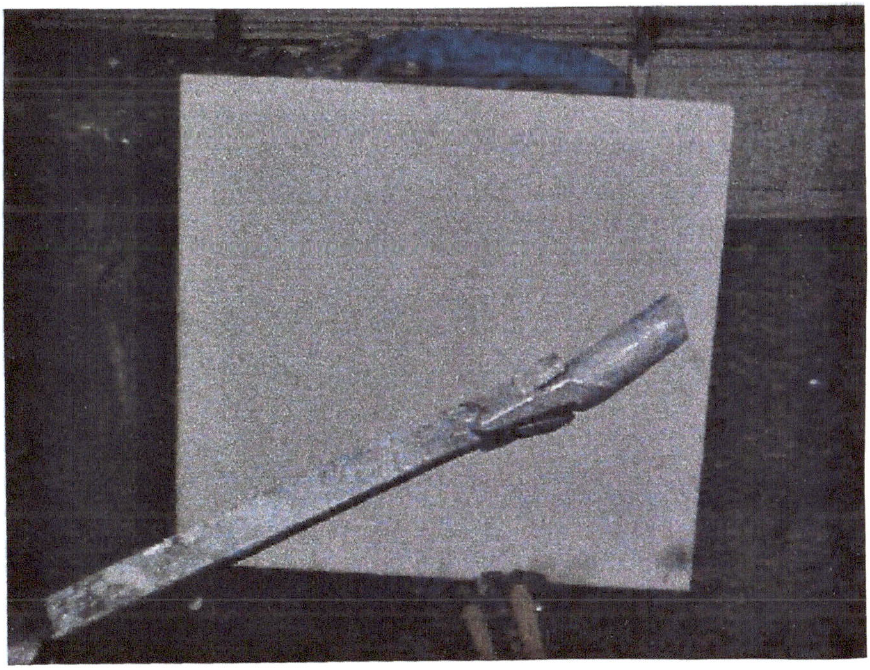

Picture 376
Here our split ends have been closed to touch their mate. The closure does not have to seem a true lock but just snug. The smith should now be able to turn and bring this forging parallel to the ground without it coming apart. Next is to flux and bring the union to a yellow welding heat. Hammer all sides to secure the welded union.

THE CLEFT WELD

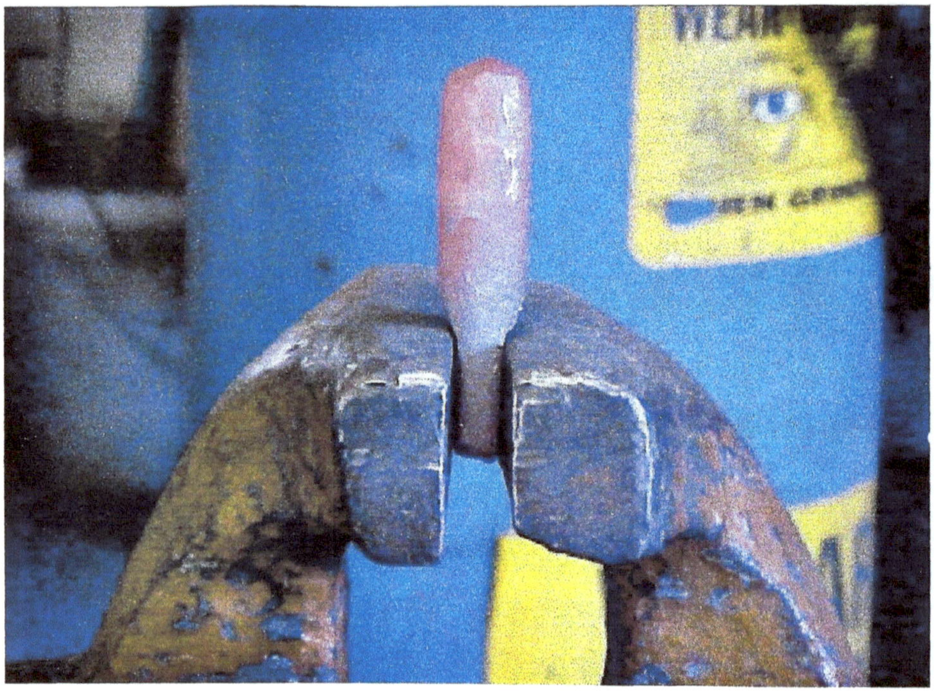

Picture 377
One means of achieving this weld by ourselves is as shown. A one inch round piece of sucker rod has been flattened on one end to create shoulders. A near side anvil edge heel blow does this. We now have a shoulder to forge against and a flat section for our vise to hold.

Picture 378
With a narrow hot chisel, the steel can be split along the vise as shown or across the vise if desired.

Picture 379
Across the box leg vise cleft split set up.

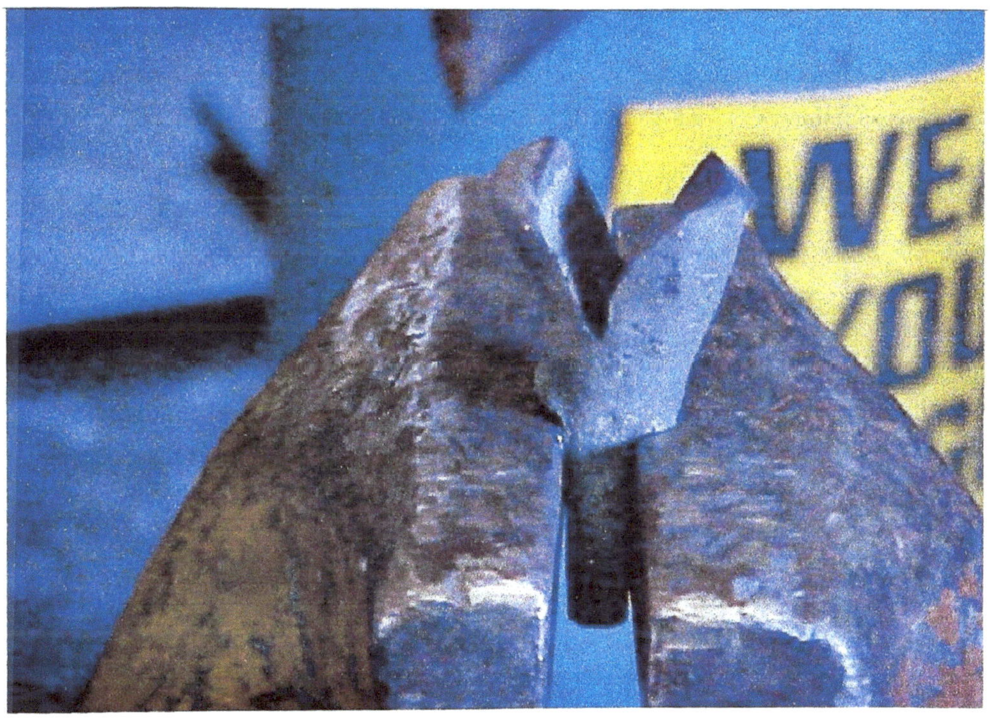

Picture 380
The finished cleft split is now ready to accept its male counter part-flat steel with a slightly (1/3) larger knob on its end. The split ends are then forged around the knob after position placement. We are then ready to weld.

Picture 381
After a cleft is cut it sometimes is necessary to part the halves additionally. If a "V" top tool is **not** handy while it's in the box leg vise, using an anvil corner may suffice.

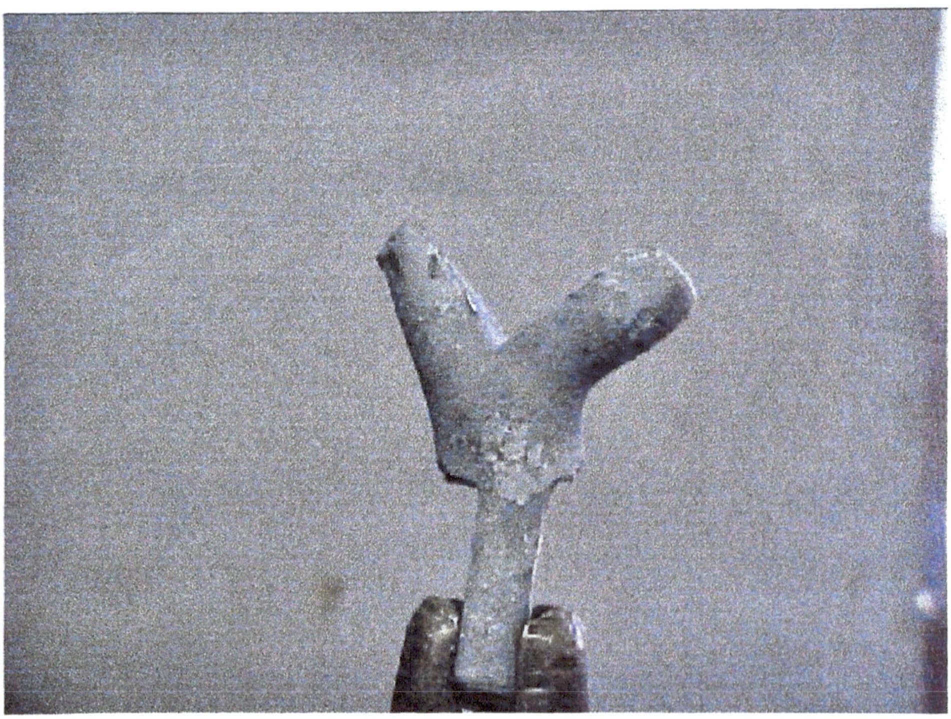

Pictrue 382
The cleft can be driven to 90 degrees open and beyond.

THE CLEFT WELD BALL

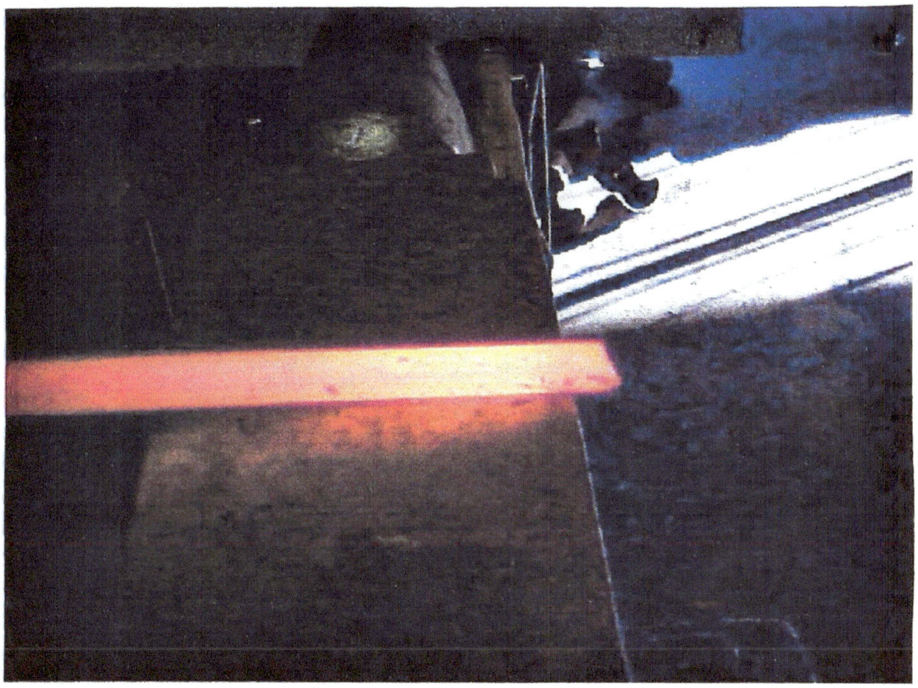

Picture 383
For this half of the weld union, we are starting small and making the ball end larger than its original stock thickness. The amount of steel hanging beyond the far anvil edge is bent over to create a near 90 degree angle.

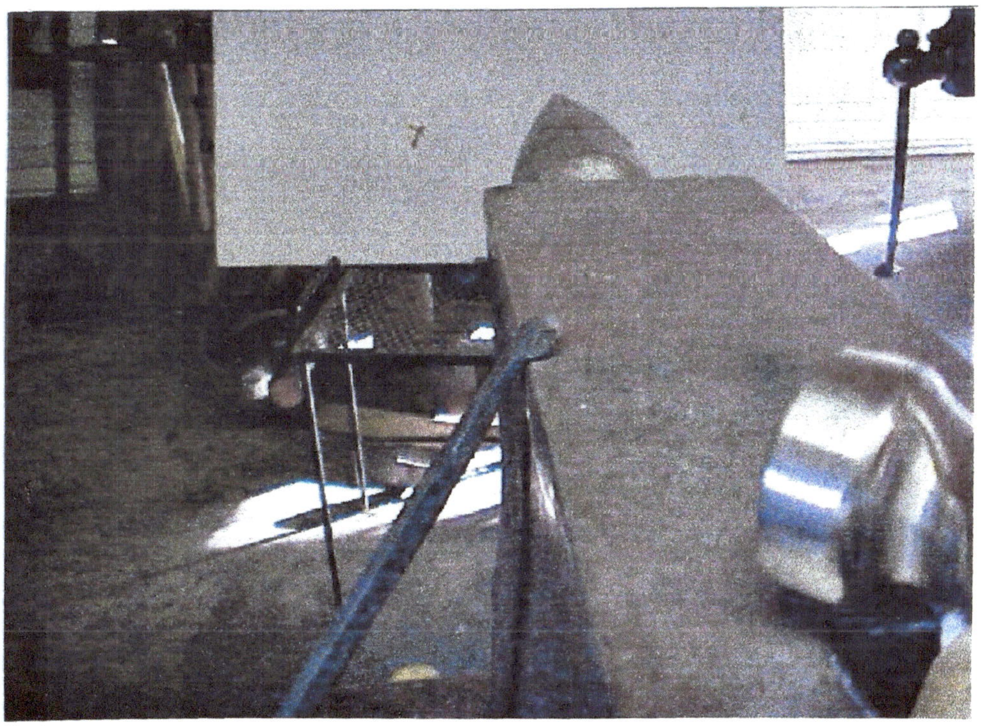

Picture 384
We then come to our near edge and begin packing the end, rolling our steel upward and downward while the hammer hits the same location. The steel is also turned over to enhance uniformity.

Picture 385
The hammer blow is a near edge half face push.

Picture 386
The finished ball does not have to be perfectly round; we are just after the shoulder it creates under the ball.

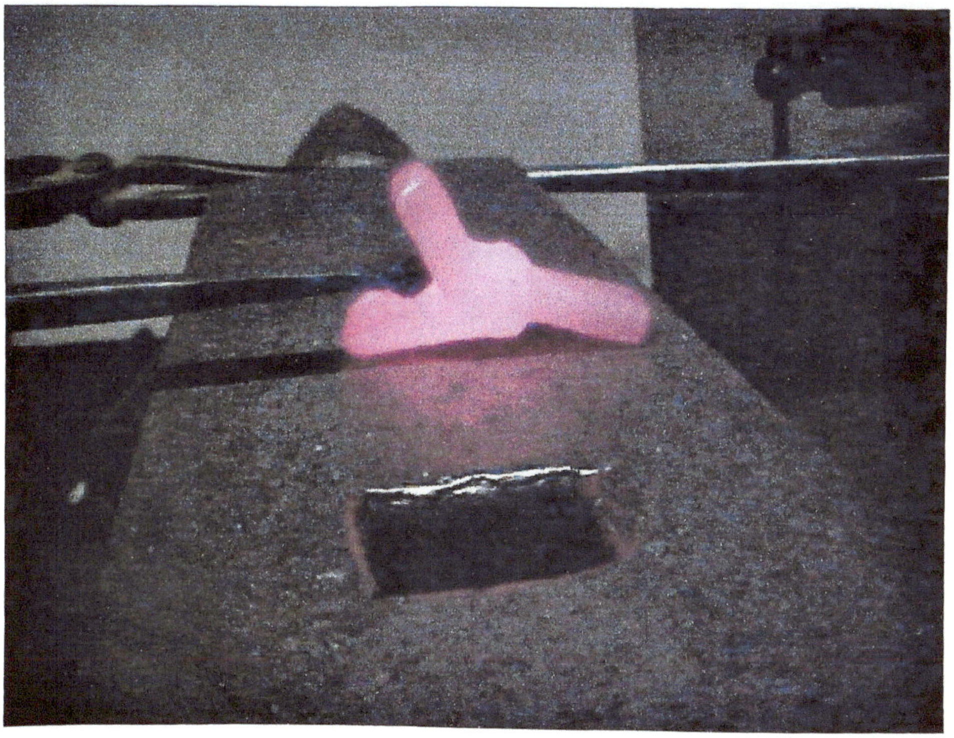

Picture 387
Checking your ball size to cleft split ratio is helpful. Is the cleft too shallow? Is the ball to large? Should we raise a bridge or lower the ocean? The fit shown here is just about right.

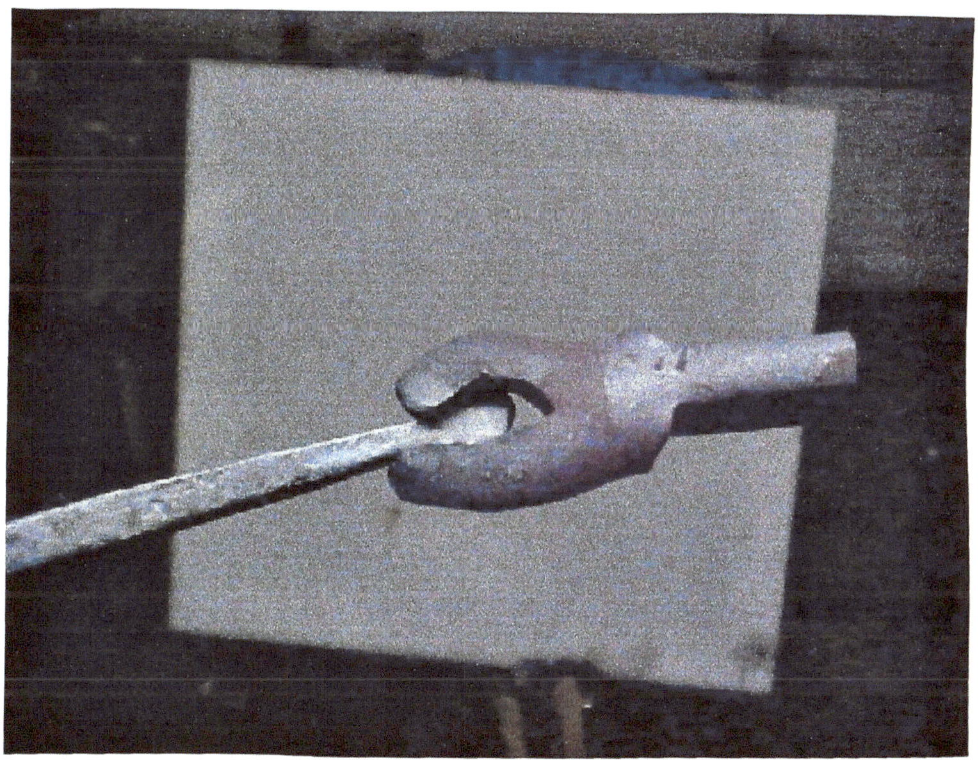

Picture 388
The cleft weld fit shown here has a little too much air space in between. Modifying the ball to a tear drop profile is all that is needed…

THE BRAND RE-CREATION

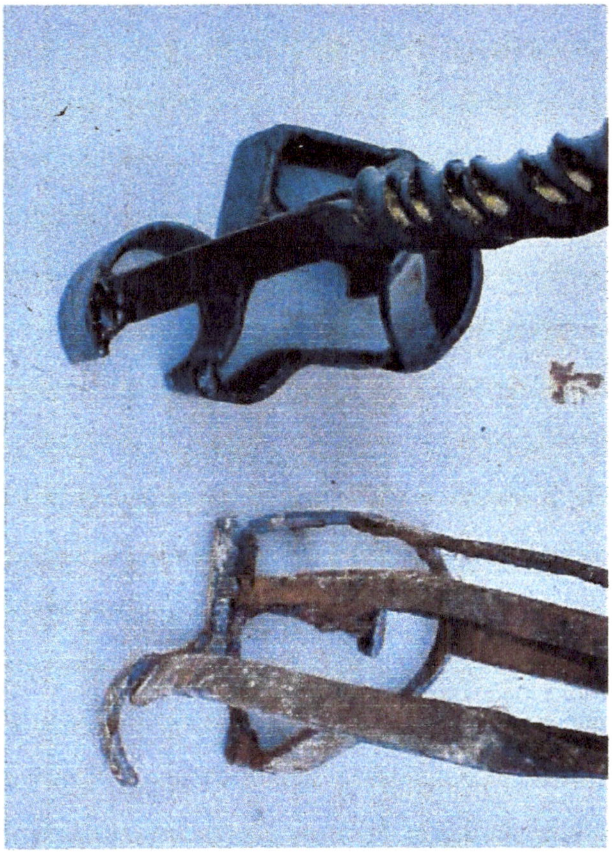

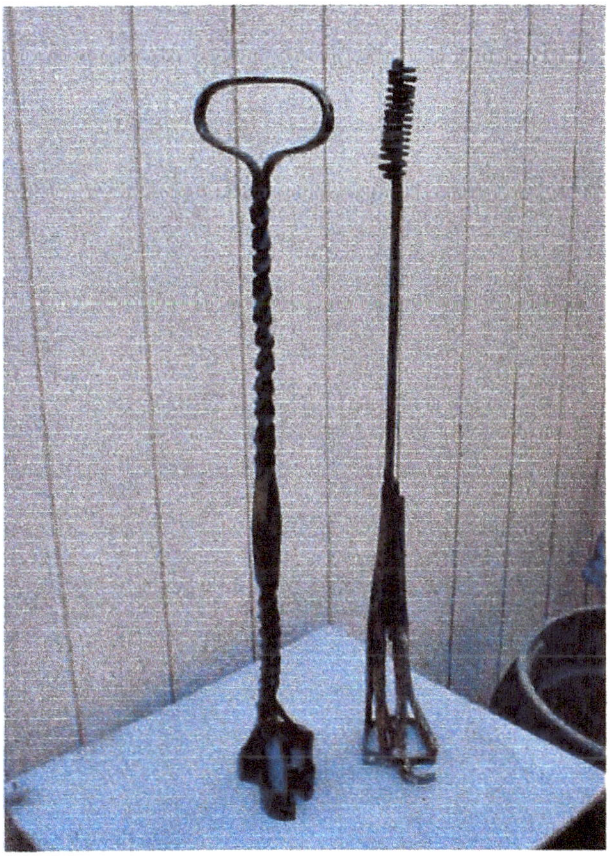

BRAND RE-CREATION

Brands of all sizes and shapes have been exhibited in our shop. Most are registered and have specific measurements they must encompass when finished and made ready to use. Some are of a ranch trademark and/ or brand, but made larger or smaller than their registered livestock version. These are to be used to heat brand fence posts, company buildings, saddles, tack, pack boxes, riding/haying chaps, etc.

Shown here is a brand that was created 20 + years ago for an elementary school. This was fabricated as a teaching aide to help instructors demonstrate to youngsters the proper use of a brand. It was heated in a firewood fire and seared short pieces of lumber as souvenirs. The original is at right of all pictures. Our recreation is on the left. The original would not have lasted for one more heating without disintegrating. Liberties were taken to provide greater thickness and width to the heat sinks and braces, 5/16 x 1 inch. The entire shaft and handle was made of one piece of ¼ x 7/8 inch stock 8 feet long, doubled over flat wise and another piece of ¼ x 7/8, 30 inches long sandwiched in between.

Hot twisting two sections as shown was done in a secured box leg vise with a crescent type wrench. The heat transfer radiator in our lower handle is a segment that was not twisted but upset to provide airspace in between the steel layers. The handle and attachment braces were parted and shaped on the anvil horn, all by hand. The heat sinks and handle were joined with a torch weld (oxy/acetylene) mild steel rod. In the old days the joinery method was jump welding or holes punched for steel riveting. Some old brands have heat sinks and handle made of all one piece. It's O.K. to work with what you've got, it must only be functional. Just be sure to maintain a sturdy and level working surface. This will reduce the amount of time needed to leave a true and crisp brand impression of flat surfaces. The rocking motion required for livestock is also reduced, especially when large brands are made. Brands can be leveled nicely after finished. Simply heat in the forge and use the anvil to make it flat with your hand hammer (cross pein, rounding, ball pein). A bench grinder can also help complete the fine tuning after it cools off. Grind down the high spots.

When cooling a brand we do not plunge it into water. This is sure to twist or tweak the brands symbols and letters. The handle may not even remain perpendicular to the heat sinks. Air cooling is best here. All pieces used were of A36 mild steel which does well with the incessant heating and cooling that is mandatory with use. If a medium carbon type steel is desired or freeze brand is to be fabricated, a trip to the steel supply and conference with the clerk may be required.

WHEN HAMMER HEADS WON'T HOLD STILL

There have been cases where our larger swing tools refuse to stay securely snug on a wooden handle. Whether initially installing the handle to a through eye which has a generous hourglass shape or one which has been deprived, some measures can be studied. These measures pertain to the handle rather than our tool head in this segment of focus.

The anomaly occurs for us mainly in the larger tool heads, three pounds and heavier. Between climate conditions (arid summer, freezing winters) the wood fibers tend to shrink. Even kiln dried wood aged for years will show just enough shrinkage a week after installation to provide undesirable shifting of a hammer head during use. This shifting descends into slop on our up and/ or down stroke of the swing. Now the tool eye is abraiding the through eye handle wood as a dull wood plane and sandpaper. Miss hit blows are almost destined to occur more and more.

Depending on your ecological climate and swinging tool eye configuration, this problem can be held at bay for years or eliminated. One of our solutions for this is as follows.

Picture 393
A prominent shoulder is left on nearly all swinging tool handles. The wooden handle is rasped down and left coarse. No finish filing or sanding. A section of racing bicycle inner tube is cut. These are usually of better quality; density, memory, durability. A used one is O.K. for we will really use it now!

Picture 394
Some will have external ribbing as shown.
Just right…

Picture 395
The tube internal diameter should fit over the through wood without stretching out. For wood that was rasped a little too much, double over the tube rubber. If the rubber appears to have the perfect density and desirable durability but is thin of cross section, it is also doubled.

Picture 396
Otherwise, single thickness with tube ribs to finish in the eye is great.

Picture 397
The lubricant used to help our handle slip into the eye without damaging the rubber excessively is <u>water</u>.

Other lubricants can be used for this as long as they dry or evaporate eventually. Boiled or raw linseed oil is one such type. The quicker it dries, the sooner handle top wedges can be installed.

Picture 398
A box leg vice was used as a back up tool to drive this handle on. Note the jaw width which allows for handle passage. The saddle tool will also work for this phase.

Picture 399
This is what the handle looks like after a portion exits our hammer eye. This is about idea,l for the excess rubber will recoil as we trim it to expose the wood wedge slit.

The exposed rubber beneath <u>is</u> all trimmed away for a clean appearance.

Picture 400
The exposed rubber above which protruded beyond our tool head <u>is not</u> all removed. This rubber will help create a wedge to inhibit our tool from sliding up the handle. With practice the smith will learn how much rubber to take and how much to leave.

Picture 401
When the water lube is visibly dry our wedge slit is parted with a hack saw. Making all more visible will save wooden wedges from destruction.

Picture 402
A "V" cleft contour is now worked into our wood. The wedge slit is further parted by rasping with the edge of a horse rasp. Minimal wood removal should occur and the progress is swift, a very important concept in a shop that depends on many hand hammers! A coarse mill file will work for this as well.

Picture 403

Many handles are installed to tool heads with no adhesive what so ever. In a perfect environment without changing of seasons and minimal work load requirements this may suffice. Complex means of compressing handles into tool eyes with hydraulic, air or drop hammer machinery power may eliminate some of the effort requirements needed to make a secure union. But no less attention to detail here can be observed. Simple cognitive planning and preparation will enable basic hand tools to perform a good and lasting job.

Our adhesive material is now mixed and matched to hold wood as well as steel. The material used should dry to firm / hard but retain some sense of pliability. Durable yet flexible. If there is some uncertainty as to inherent properties when dry, a sample plot can be tested. Mix up a small sample on a piece of cardboard and allow drying time. If it sticks well to cardboard, wood adhesion is almost certain. Try another small sample batch on a piece of remnant steel. How does it stick? Will it hold to your steel wedge? Some experimentation may be needed to learn what good adhesion to these materials looks like. But at least now we are looking for that!

Testing the adhesive before implementation may prevent a ruined handle and starting over.

Picture 404
Our final installation steps are to mix up a suitable adhesive. Most often a quick drying two part epoxy, (5-20 minute varieties). These are applied first to our wooden wedge. Many adhesive substitutions have been made here successfully- dental bonding cement, acrylic hoof repair compound, even shoe sole repair glue. Find one that works for your environment and stick with it. The search shouldn't take too long…

Picture 405
The adhesive is also applied to our steel wedge before and after installation. Remember to wear eye safety protection at all times. Let dry and continue to finish.

THE BLADE PREAMBLE

THE BLADE PREAMBLE

Over many years the number of blades forged, finished and used to extinction in our shop cannot be counted. They are one of the items forged here that encompass such a wide variety of shapes, sizes and uses, an individual cannot have enough of them. This is almost on the level with hammers, almost. A tool perhaps created before during or shortly after the first hammers of primitive ages, can have both great functional and aesthetic values. The consumer demands for a good knife have not been viewed to fade. Some have proven more desirable and precious than others, but all have found their place.

The types of blades and billets which can be forged proficiently are innumerable. The blade smiths and knife makers are assuredly able to develop a market share of that alone. Some blades may require months or years to be fully understood and constructed. Then a life time to master. This is in regards to folded and welded blades. Our requests for blades have included the following:

4" hook less hoof knife - Horseshoeing
3-5" hoof knives - Horseshoeing
5 - 7" hoof knives - Draft horseshoeing
5" loop knives – Horseshoeing
10-12"cutlery knives – Fruit and vegetable
6-10 ½" hunting knives
7-8" throwing knives
5-12" skinning knives
8-12" butchers cleavers, choppers
30-36" broad swords – unsharpened for renaissance battle practice
36-42" broad swords – armor display
8-9" axes, throwing and firewood splitting, have hole for wooden handle and weighed.
26-32" Samurai swords, short wagasachi
30-37" Samurai swords, long katana

These measurements are of the blade only. The tang lengths for handle attachment are added to these numbers. The thicknesses and widths of each blade are decided upon as each is described for forging. The type of steel selected for each is varied. It has been decided through trial and error experience that no one alloy will make them all correctly. The shape, profile of width and thickness contour has much to do with what a blade can do in use. The type of steel and heat treat process has done more to determine what a blade cannot do. Bend, pry, flex, impact resistance, edge holding capability, toughness against chipping and cracking, wear resistance when in use must all be prepared for with the heat treatment of the end product blade desired in mind.

Beyond the myriad of blade varieties and style, there is a need for proportion suitability. To whittle and carve wood efficiently a knife should be one size, to hammer a blade through joint cartilage dressing out a hunting catch it should be another. Overall the blade should <u>feel</u> right. This is different from one person to the next. Right or left handed individuals. Large or small hands will have a preference. Male or female appendages of varied degrees of dexterity and strength must use the implement safely. This brings importance to bear not only on the blade but the handle as well. The handle and finishing will be the current subject of focus.

Hoof knives for trimming horses hooves. These blades are all from 440c stainless steel, annealed after forging. Oil hardened and tempered to a straw color for a harder edge. Ergonomically correct handles were carved and sanded to fit one farrier's hands; your author. The starting steel blank dimensions were 3/16 x 1 x 5 ½". The handle screw holes were hot punched at a red forging heat, just as a rivet hole for our tong. Hickory and Osage orange wood was the handle material. Epoxy glue spread on the blade tang before union helps secure a tight contact.

The tool pictured at bottom is a hoof pick. Forged of mild steel 5/16 x 1 x 9" blank. No heat treatment. The twisted "0" ring at one end helps keep it exposed when dropped in any barn soil. The pointed end shows the profile which develops from years of use. Design for these should also be ergonomically correct. Numerous shapes and sizes of these should be present in any blacksmith shop, even if horses are not shod there regularly a rock can be pulled from your boot tread with this tool…

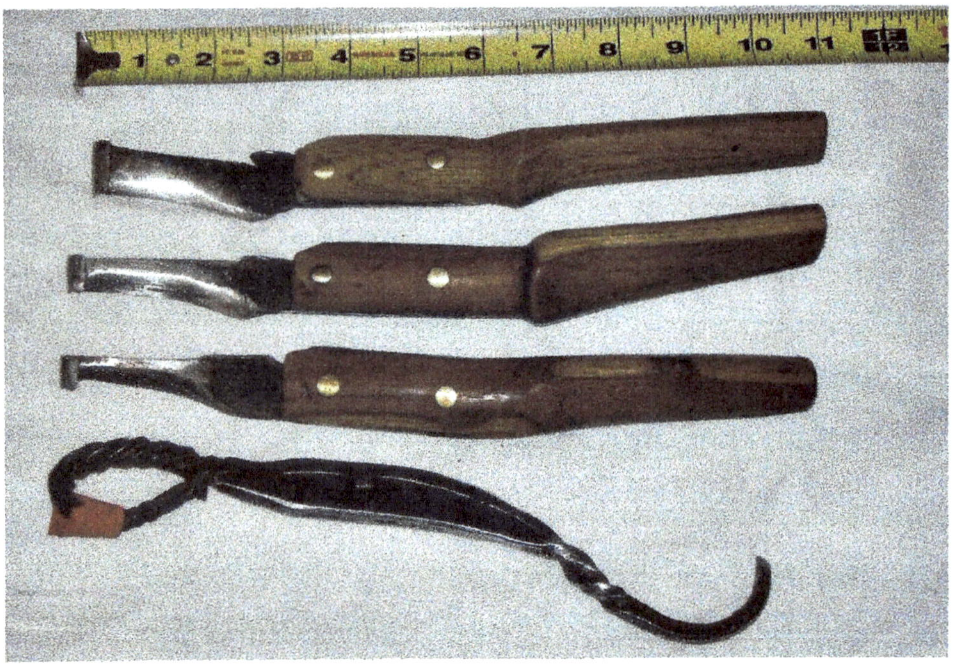

Picture 406

Picture 407

The hoof knives have been forged to provide a mousselet clip. As it applies here, we refer to material left unhammered during blade formation. After the blade is forged nearly to its final shape the mousselet reservoir is hammer drawn away and up from the blade. The material is lastly hammered into a position to remain perpendicular to the entire blade. The function for this clip is as a thumb pusher. This added feature helps the farrier utilize both hands more effectively. Especially when prepping hard dry hooves.

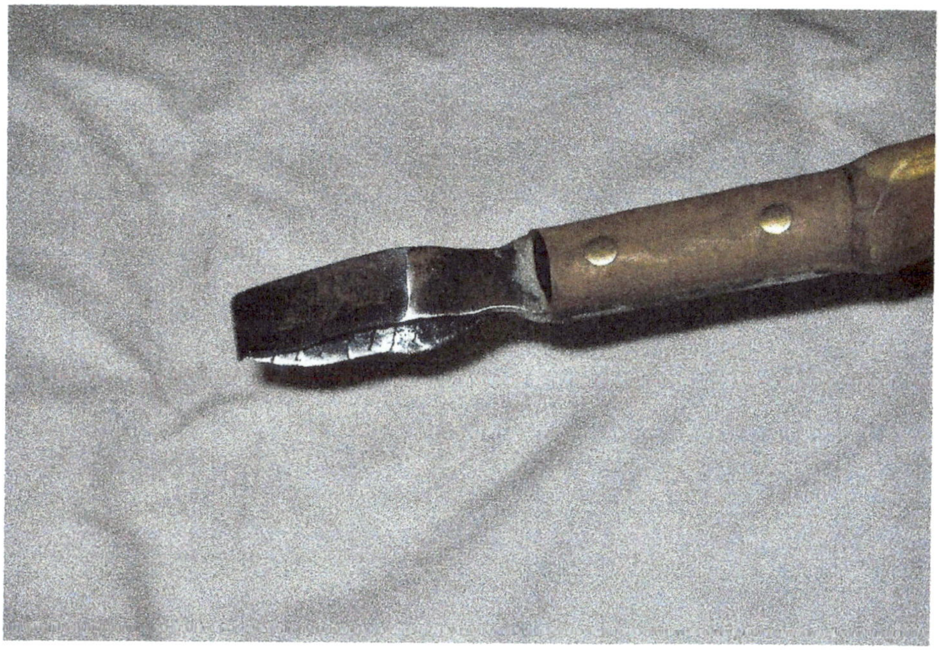

Picture 408
The loop knife
side view

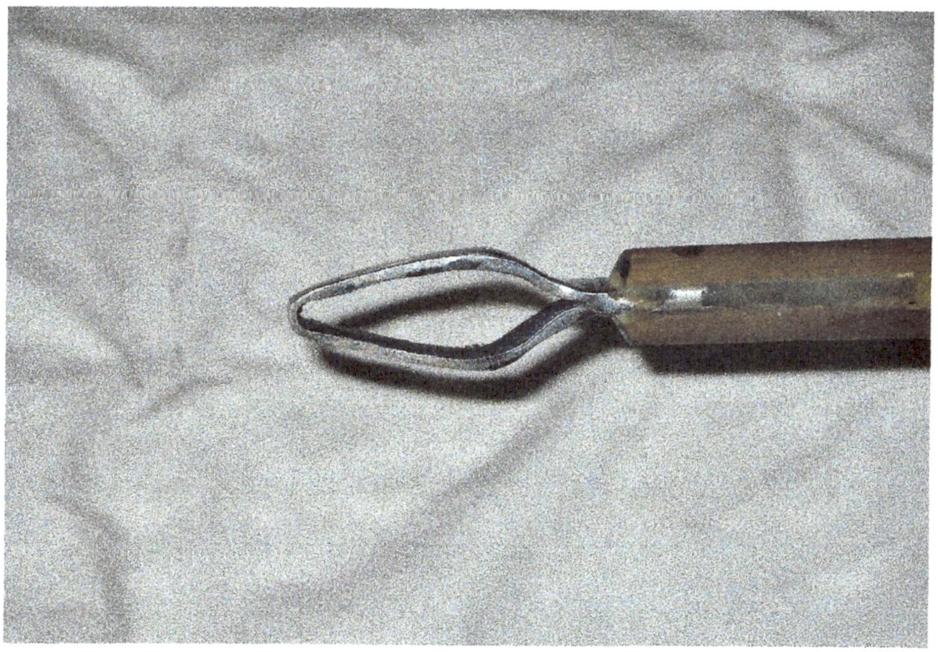

Picture 409
The loop knife
back view

The Farriers/Wood carvers Loop Knife

- Blade materials (Steel) : 440C stainless, A2, old files, hoof rasps, planer blades
- Screws : Double blind nuts (brass)
- Handle materials (Wood) : Osage orange, hickory, ash, apple stems, eucalyptus
- Glue : Epoxy (5-20 minute type)

Steel Dimensions : 3/16" x 1" x 12" (cut length to suit needs)
3/16" x 3/4" x 12"

Wood Dimensions : 1 1/2" x 8" extra thickness for ergonomic shaping when finished

Overall length : 11" when finished

Top view

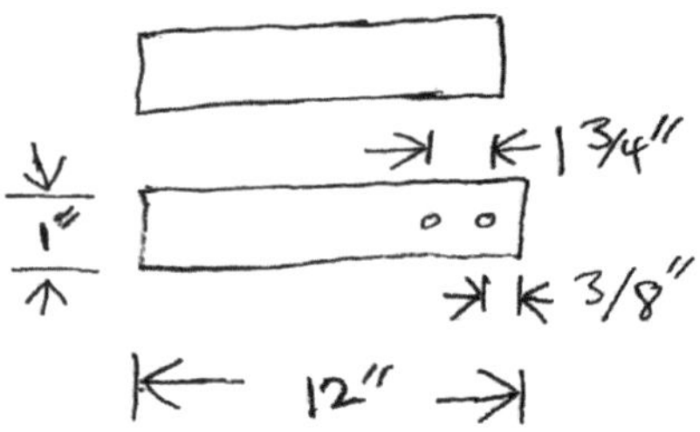

Double over flatwise

Side view

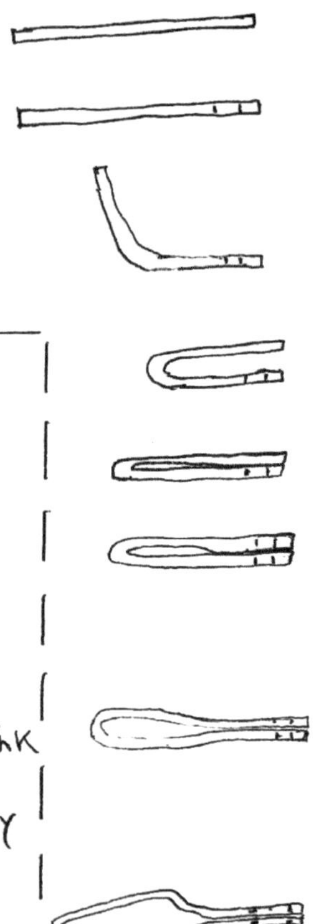

- All done at a cherry → orange heat
- Tang screw holes hot punched one end
- Hot punch second pair of screw holes using the first as guides
- part edge portion with hot chisel
- Tang held together after heating (clamp, vise, vise-grip)
- Drift punch driven through edge section
- Can be opened on anvil horn

To Finish
- Anneal
- Rough grind inner and outer edges.
- Harden (oil) slow rising heat to non-magnetic, quench
- Temper: flash off oil
- Regrind/dress
- Prep Handle: slot, align, drill holes, counter sink
- Install blade: with epoxy and screws, let dry
- Rasp, Grind, hand sand

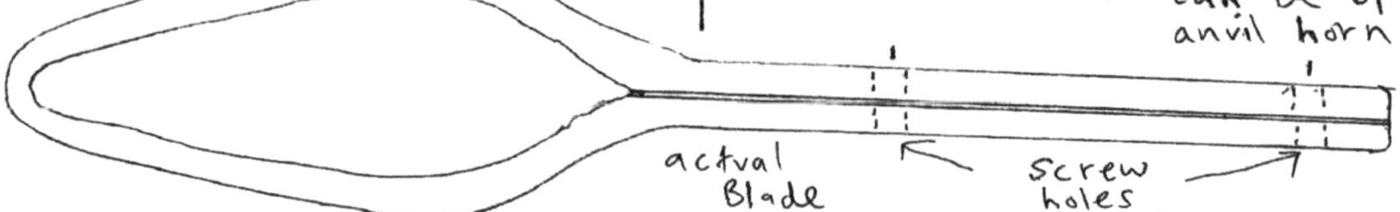

Picture 410
The samurai style sword and scabbard. Forged of 5160 truck leaf spring remnant for technique practice. It's easy to forget steps if they are not practiced from time to time. Dry fitted and ready for a final grind and polish.

Picture 411
Using the built up box style, our sword in a cherry and walnut wood full scabbard. Handle material is Peo Ferro, iron wood surrounding a ¾ tang. The handle is two slabs, each chiseled to accept one half of our tang thickness…
Three quarter tang means the steel inside and covered by wood encompasses that much (3/4) internal length of the handle.

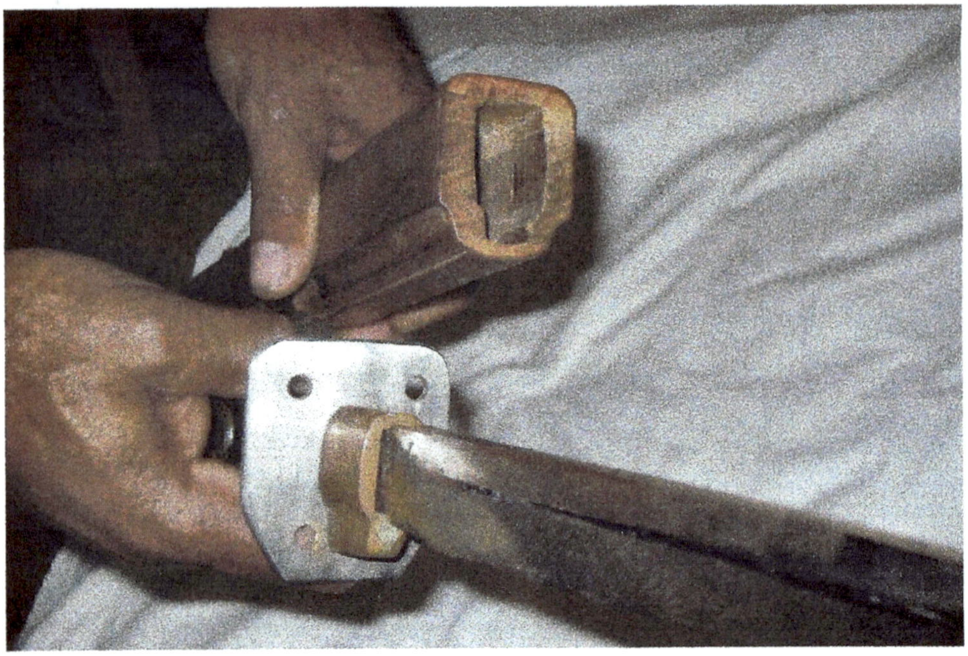

Picture 412
Entry view of the scabbard at top. Carved to accept only this blade and tsuba (copper and nickel silver guard retainer/scabbard lock).

Stainless steel guard with decorative holes. These holes also serve to lighten the overall project. Light weight is good for these blades will cut air when wielded.

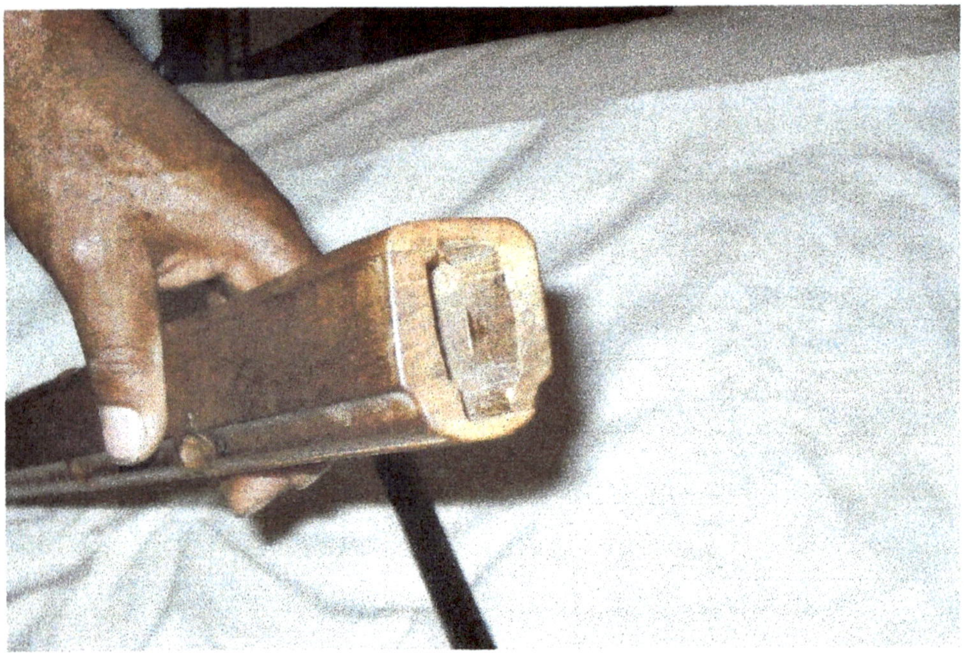

Picture 413
External wood routing adds much to the scabbard amenities. Smooth sanding, tongue oil and saddle waxed exterior only.

HOW OUR KNIFE WAS FINISHED

HOW OUR KNIFE WAS FINISHED

The knives produced in our shop are nearly all of the forged variety. They are made by shaping the steel in some form after fire heating to enable malleability without excessive force. The advent of better quality steel and increased availability makes it possible to merely grind and machine a blade into any desired shape. A problem which occurs with the ground to shape blade is the interrupted grain structure, the internal constitution which results from complete stock removal. This method may not present any deleterious effects whatsoever if attempted diligently without excessive heat development. Cooling while grinding and heat treating within the manufacturing parameters can provide satisfactory results.

For our blades which receive even minimal forging to shape, the handle pinning holes are hot punched. These blades are built to last and most refuse drilling.

Shown in this chapter is a recent hunting knife project, forged after the welded and folded method. The pictures viewed are all of the same knife, directly after the forging stages to finished and ready for the show case. This particular knife took an overall 1st place blue ribbon of the metal art class of El Dorado County Fair 2004 Placerville, California.

Materials were: 2/3 Horse rasp
3/16 x 2 x 8" L6 saw blade remnant
(2)¼ x 2 x 10" mild steel

Four layers to start, one piece mild steel is fire welded to each carbon steel piece one inch down each billet. They were stacked flat wise with all edges flush. The four separate pieces are now two thicker ones. They too are now stacked together flat wise all atop each other fluxed with borax. The borax is allowed to bubble up and boil its moisture out, leaving only a glassy clear film. The flux now called calcined and ready to help with another fire weld. The steel is stacked so that mild steel touches carbon steel, the best way to get the first fire weld. Continue welding as little or as much at a time as can be hammered with good adhesion. These billet heats are yellow-white in color but not dripping hot.

With each folded and welded blade the smith needs to make a mental note as to the progress made from each hammer blow for a given billet thickness. When many welding heats are used the steel will loose carbon and structural integrity regardless of the steel type. After the steel has endured some of these heats, a difference will be noticed in how it moves. When the welds are secure but the billet steel moves too easily for forging, it is giving up and may not finish as a hard blade after heat treatment.

Picture 414

Picture 415

When the entire piled billet is welded, we fuller cut three quarters of the way through the middle flat wise. This will enable folding the in middle flat wise. This enables folding the billet easily. All edges and length are aligned and welding continues as described. From now on, after each full billet weld is completed, the piece is then drawn for length. It is much easier to hand hammer this way; before the thickness becomes too dense to forge.

At some point one end may develop tidier than the other. Mentally keep track of which end is which. The best end should be our blade tip, the other will be the tang.

Our blade is gaining layers exponentially throughout this process. Some intricate blades of old Japanese, Damascus, Persian, Viking and many others placed untold effort on pattern welding and design in this segment. We are just covering the tip of an iceberg here to help student understanding of the blade shown.

The layers were increased from 4 - 8 -16 – 32 - 64 – 128. Our blades are turned up on edge to be forged into their final shape. The weld lines which were on the sides of billet are turned toward the sky and hammered down to create our blade shape. The weld lines are now exposed on the sides of our blade. They are exposed for all to see and test the quality of the welds. Inspecting things as the forging continues, there is still time for another welding heat if needed...

Hot punched tang holes were made after the shape became complete. Our name stamp was also done at a red forging heat. All was checked for straightness. A little steel was left for final dress grinding.

The heat treat process for us begins immediately after the forging is completed.

-- Anneal, rough grind- 60 grit belt
-- Guard brazing- Nickel silver + borax + flux
-- Harden- Slow rising heat to non-magnetic plunged vertically into warm water.
   Tang and guard not showing color.
-- Re buff- deburring wheel dress
-- Temper- Straw-brown color

With each folded and welded blade the smith needs to make a mental note as to the progress made from each hammer blow for a given billet thickness. When many welding heats are used the steel will loose carbon and structural integrity regardless of the steel type. After the steel has endured some of these heats a difference will be noticed in how it moves. When the welds are secure but the billet steel moves too easily for forging, it is giving up and may not finish as a hard blade after heat treatment.

Basics of our folded and fire welded billets.

eg. Hunting Knife blade

Materials: Tool steel – L6 (saw blades)
5160 (coil springs)
Files (Horse, mill bastards)

mild steel – A36 (remnants)

Nickel silver – round & fluxed to braze guards (non-build up type)
Flats to add to laminate layers

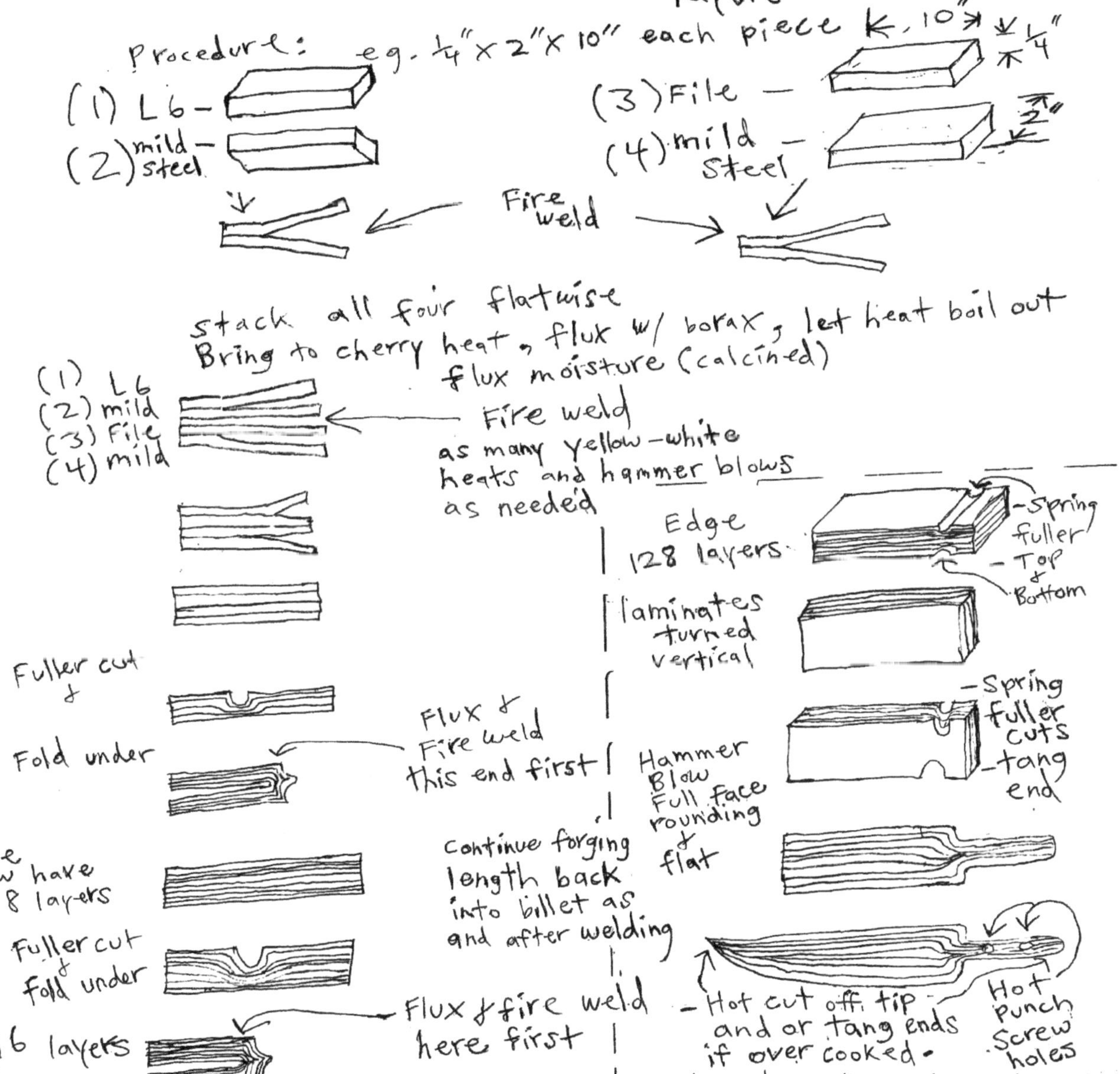

– Anneal, grind, harden, deburr, temper, braze on guard, polish, etch, sharpen

Continue folding and welding. Between 64–128 layers steel may become softer. Carbon loss, watch heats...

GUARD CONSTRUCTION

The style, size and shape of a blade guard are as varied as the materials from which they are made. The smith can also decide if one should even be installed. Will the guard have a functional aspect or be merely decorative? Our guards are typically of non-ferrous materials; stainless steel, copper, brass. Most are attached to the blades to remain permanently. The exceptions are Japanese swords and some broadswords which have removable guards for cleaning and shimmed for tightness as they age. Guards with heritage engraving are easily changed should a blade change ownership.

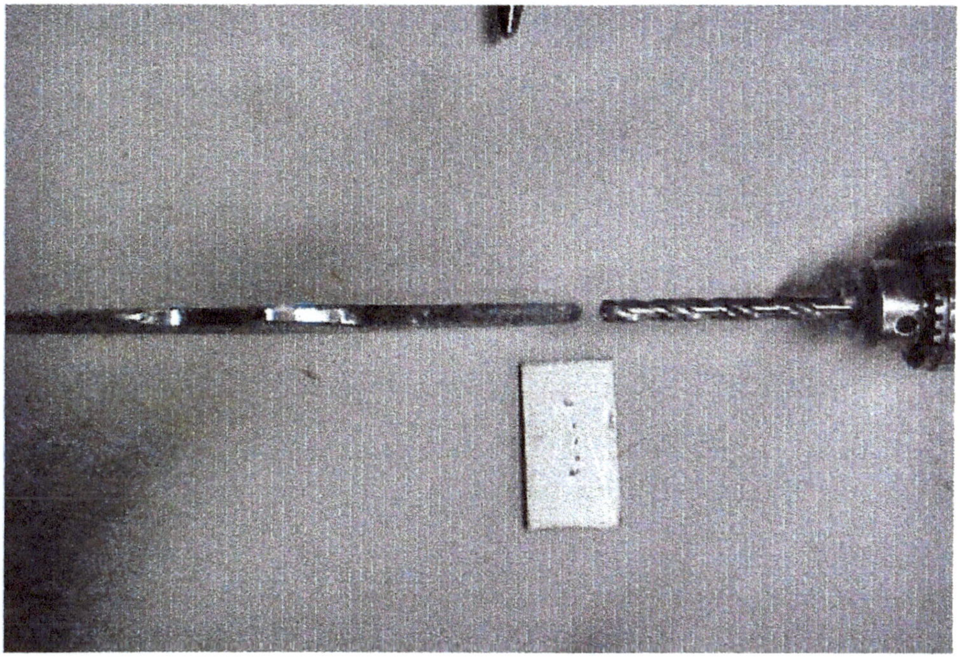

Picture 416
A quarter inch thick stainless steel guard. Hack saw cut to size. A drill bit is selected to match the tang thickness.

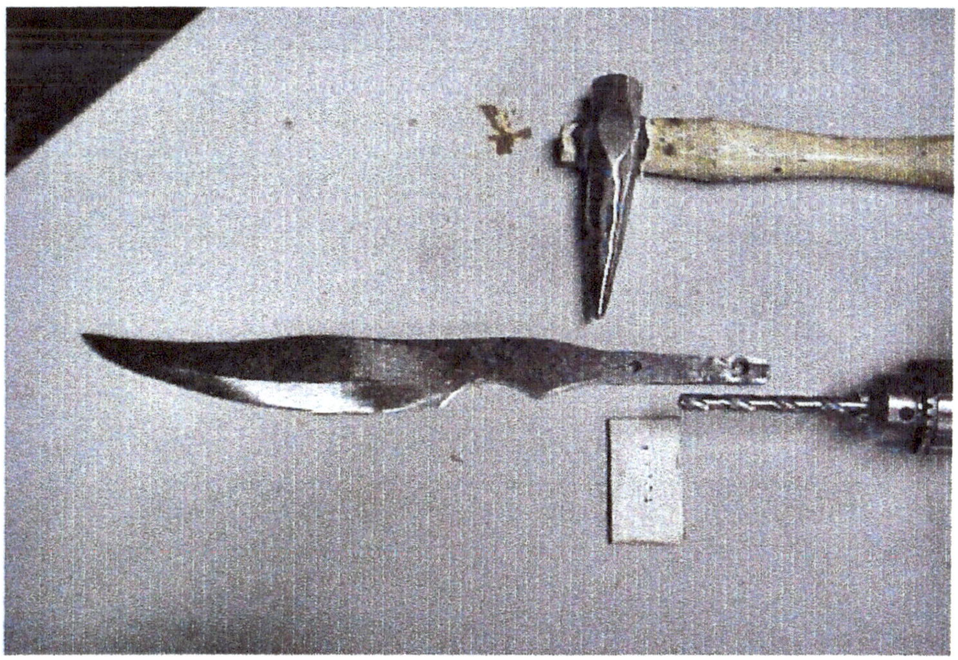

Picture 417
Center punch marks made on the stainless guard in succession. Multiple holes are drilled depending on the desired resting place for the guard.

Picture 418
The guard to be secured in our box leg vise is protected by a heavy leather wrap. Stainless steel is softer and will otherwise scar.

Picture 419
Our through hole was hand drilled.

Picture 420
Rounded edge mill filing commences to acquire a good hole shape and snug fit to our blade tang.

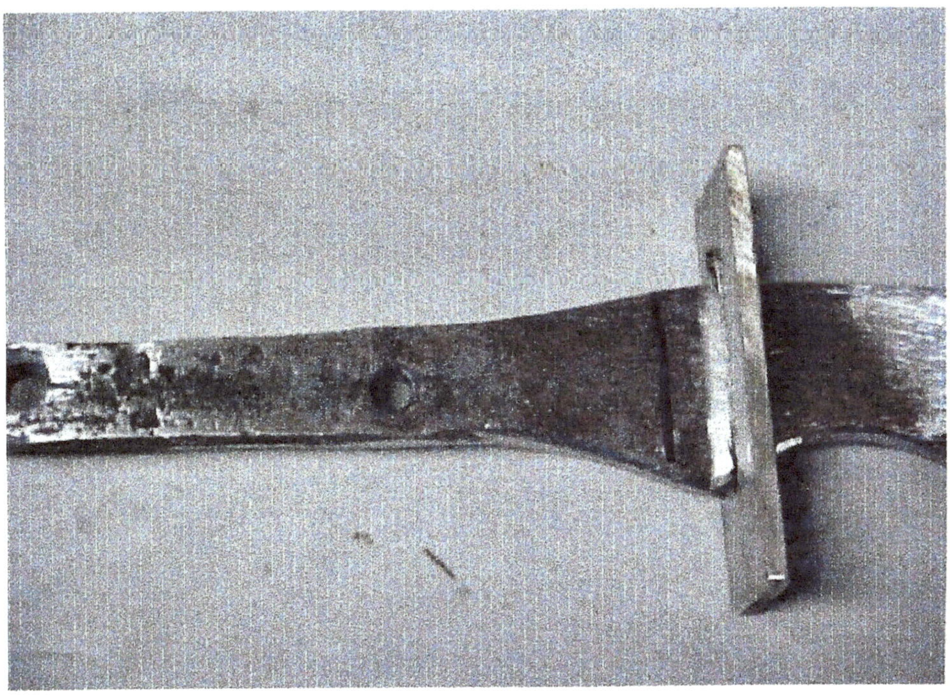

Picture 421
A magic marker line is made on the blade as a reference point. Filing continues until the guard fits along this mark.

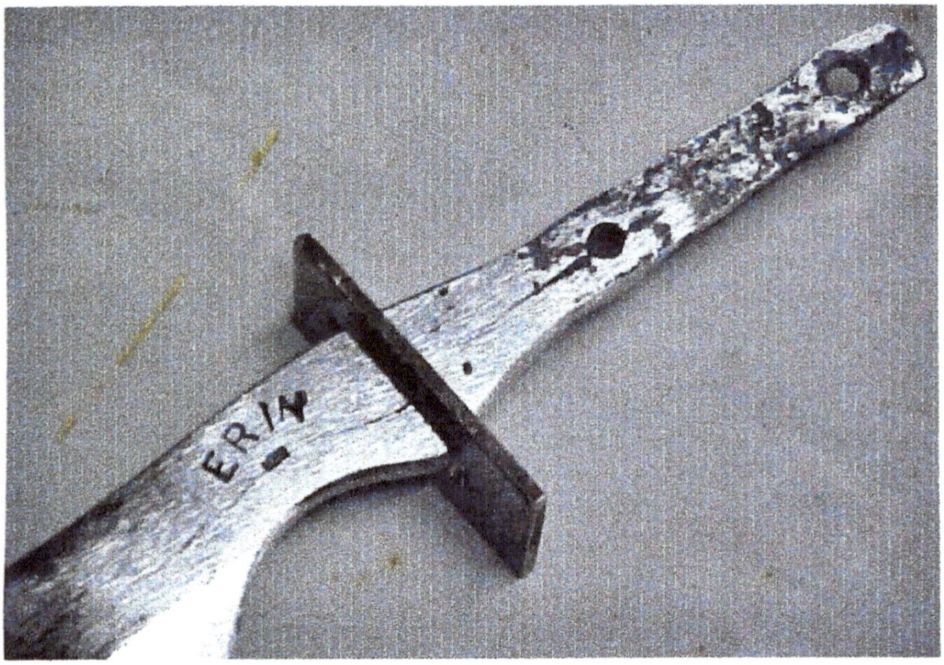

Picture 422
Here our guard fits well when moved into position over the tang. Some blades with a wide tang will prove easier to fit and slide a guard on over and past the point. The tang (at right) is <u>left</u> coarse for handle glue adhesion.

Picture 423
Our guard and blade are suspended on and in the box leg vise. All is now ready for brazing with the oxygen/ acetylene set up.

Picture 424
The guard is brazed from both sides. Now it won't come or go with hard service. Nickel silver filler rod was used for this particular example. Many other brazing materials can be used here, even silver solder at a lower temperature. Soldering rod will run out if the blade still needs to be hardened and tempered.

A rough grind is done on the fresh brazing. The blade is now heated to critical (non-magnetic), quenched in warm water, deburred and reheated to show temper colors. Purple/blue (spine), brown (mid-rib), straw (edge)is the goal.

Picture 425
The entire blade is again water quenched to stop the process. Review of the results is usually rewarding at this point.

HANDLE BOLT SELECTION

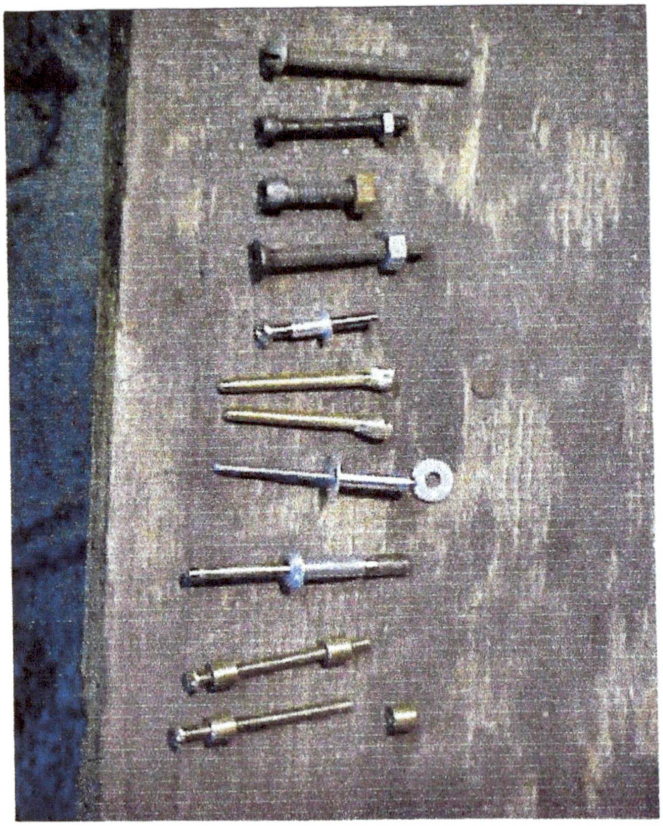

- Common machine screw
- Grade 8 Allen head bolt & nut
- Grade 8 Allen head bolt & brass nut
- Counter sinking Allen head bolt & nut
- Phillips screw & threaded back up
- (2) steel pop rivets with brass color anodizing
- aluminum pop rivet with back up plate
- Heavy duty pull pop rivet
- (2) Double blind nuts
  and screws (brass)

Picture 426

At some point during your finishing endeavors a ready supply of suitable bolts should be inspected. Our tang holes are hot punched to fit the specific bolts to which they will be mated for life. A sample of each is kept near the anvil during blade forging. An attempt to tie in the bolt type; color, shape size and length with the handle motif is ever present.

TANG AND ANTLER HANDLE

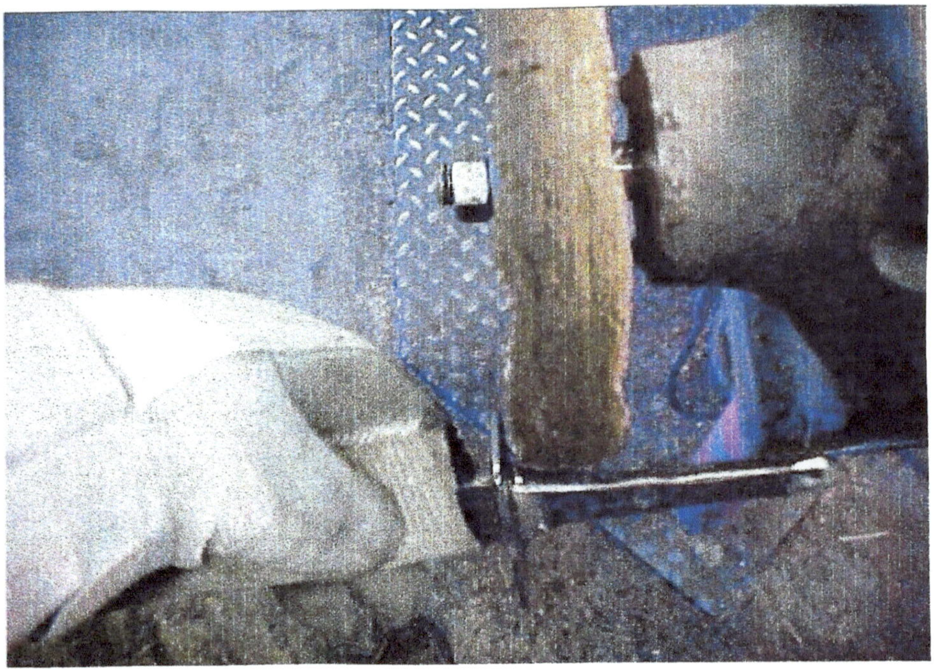

Picture 427
When tangs of various length and widths are present there may be some limitation for dressing all portions of the project. Creativity for finding a way to all nooks and crannies may keep the smith thinking.

There are many ways to the root of that evil. Trying from all angles is good, but keep safety first and foremost. Remember for smithing, we can raise the bridge or lower the ocean so to speak. We can move the project and our grinder holds still or vise versa. The project holds still and we move the grinder. Pick the right tool and means to an end...

Picture 428
The last of our nickel silver brazing is cleaned up front and back on a muslin buffing wheel loaded with 200 grit rouge.
*Note the fire hose shielding the blade from incident.

Picture 429
The handle selection is made. Stag antler left side for a right handed person. Right antler for left handed person. This is how nature grows them. We just find the best use possible. Held in our vise with a leather wrap. Hollowed out to accept the blade tang with a hand drill.

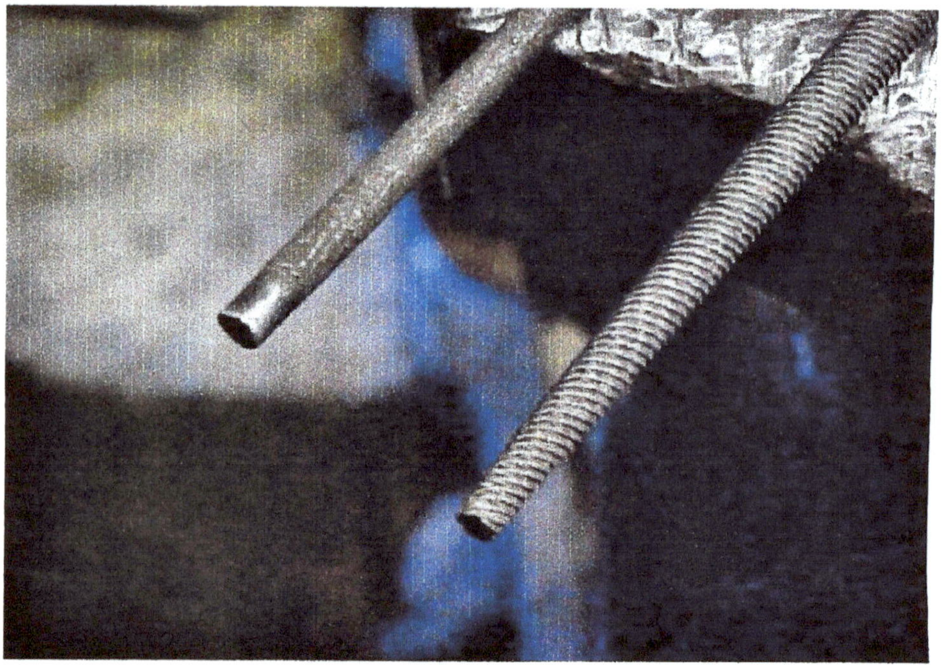

Picture 430
At left is a chain saw file and a coarse round wood file. Both are used regularly to finish hollowing work in antler handles.

Picture 431
The grinder with a 60 grit belt is used continually on steel, wood, antler and other materials. A good belt can become loaded with debris. To unclog it and restore abrasive qualities, a school gum eraser is pressed against it- an inexpensive and reliable solution.

Picture 432
Double blind nuts were selected and used to approximate the handle holes to be drilled.

Picture 433
Holes to be drilled are checked internally and marked externally. A pilot hole is run through first on each mark. Next is a bit the size of our threaded screw for passage. Lastly a bit to counter sink the nuts, only half way through the antler wall thickness.

Picture 434
The blade is now dressed to near finished. It is hollow ground with 60 and 120 grits on a small radius fixture in the belt grinder. Next to the 200 grit rouge. Then to deburring wheels and mirror finish. To keep the hollow ground gutter symmetry following the blade shape its spine is in constant contact with the grinder tool rest.

Picture 435
Leather shims are now cut, hole punched and dry fitted.

Picture 436
Occasionally a dry fit handle hole will not cooperate for the counter sunk nut portion. If it won't fit dry, then filled with epoxy glue is sure to not work. Here, off setting the countersunk hole in a drill press. A precision machinist's vise mount is holding our antler handle wrapped with leather to protect it from unnecessary scarring.

Picture 437
A good blind nut fit and countersunk hole. All is ready for final fit and glue.

Picture 438
When the handle is partially filled with glue there can be difficulty pushing the screw through both handle and tang without loading the threads. When first pushing the tang in, run a large carpenters nail in each hole to align things and clear a pathway. Remove one at a time and insert your permanent screws.

Picture 439
When both screws are in and nuts threaded on, the tightening can commence. Should the distal blind nut (farthest from the operator) insist on turning with the screw, an angled needle nose plier to hold it still will help. A hexagonal or square nut does not have this problem. But they just don't look as pretty when finished….

AN END CAP

Picture 440
To further spruce up the knife handle, an end cap can be added using skills previously introduced. A ¼" thick copper cap made from a large remnant washer.

Picture 441
Two different drill bits are used after center punching to match dry wall screw dimensions.

The pictures on the following two pages show our gradual transformation of the copper to a finished heart shaped end cap.

Picture 450
To fit screws to an asymmetrical handle, sometimes shortening one will prevent protrusion. A long screw on short corners will go through an end cap and out a side of the handle. Why let that happen?

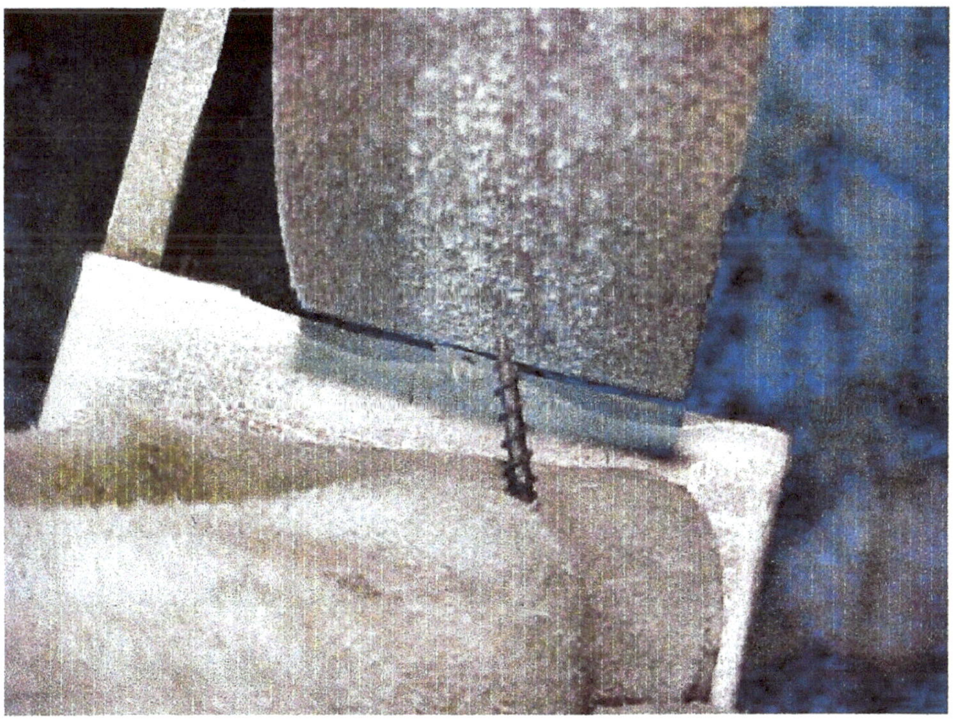

Picture 451
The end cap screw can be replaced for a shorter one or made shorter.

Picture 452
Time to add glue to the attachment surface for our end cap, provided the smith is happy with the angle it presents. There is still time to change things for aesthetic or ergonomic reasons.

Picture 453
Engraving is most easily done before the cap is affixed to the handle. This one was placed on a shop anvil and hand stamped with a date and arrow through our heart. Consider the possibilities.

Picture 454
Once all the glue has completely dried the project is soon becoming a tool. Gloves are worn from here on to keep hand salts from pre-rusting our blade, as well as for safety reasons. Safety eye wear is also in place, for certain fragments of glue, leather, antler and abrasive particles do become airborne.

Picture 455
Handle corners, edges and all surfaces are dressed for smoothl appearance and transition. Dissimilar materials will jump out in eye appeal when their surfaces are blended smoothly and evenly.

THE ETCH

The type of etch referred to in this segment pertains to controlled erosion by definition. There are numerous means of systematically deteriorating, engraving, removing or altering the surface characteristics of our steel to create the desired visual affects. We can localize our esthetics with laser technologies or broadcast the contrasts through liquid solution applications.

The method covered here is in brief acid bath solution. The etch can be of long or short duration, acidic or alkaline in solution Ph. The visual affects resulting are a product of time, solution type, solution concentration, type and combination of steel to which it is exposed. The intent is not to merely remove scale or rust, but to alter the surface characteristics of polished material which will hence delineate their beginnings and end.

Different materials respond at different rates to a common stimuli - when a billet (a knife blade in this case) containing numerous dissimilar types of steel are exposed to a common etch solution for an equal period of time, the presence of each and how they are aligned is well revealed..

Protective attire in clothes, eye glasses, latex and/or rubber gloves, face mask and good ventilation are all important safety concerns. Even our mildly acidic solutions can have long term detrimental effects from short or prolonged exposure to our body through physical contact or respiration of airborne vapors.

The faint of heart and lungs should wear a respirator if in close proximity to their acid etch solution. Many publications have been written on this subject and the harm which can result from improper handling… Do research if any uncertain questions are at hand!

For our blade, a plastic bucket of water has the blade submerged almost to the guard level. Enough fluid height is left to add the mild acid solution. The acid <u>is</u> added to the water, not water to acid as a violent reaction can occur. Our mixture was muriatic to water (swimming pool cleaner) a few ounces at a time while wearing surgical gloves in case of a splash. Always have safety eye protective goggles on.

The blade is reviewed every 5-10 minutes to check the etching progress to find our ideal darkening scheme. Another bucket is ready with baking soda and water mixture. The blade is taken from the acid bucket and lowered into the baking soda and water for neutralization. Acid should be neutralized in an alkaline solution before any safe handling can take place. The blade and entire areas which may have contacted the mild acid solution are rinsed with fresh water thoroughly. All is towel dried and viewed. If results are what the smith had in mind, continue onward. Should a darker etch be preferred, repeat the given steps. If the blade became too dark overall, neutralize it in baking soda, rinse thoroughly, deburr the entire blade, bring to a high polish or luster and re-etch.

There are many mild acid solutions readily available. Here are but a few:

Muriatic acid
Ferric chloride
Vinegar; apple cider, white

The vinegar has been used to provide unusual hue to our blades. The cider can be warmed slightly above room temperature to hasten some results. Gun bluing cold will also work, but can get messy until the user develops a system…

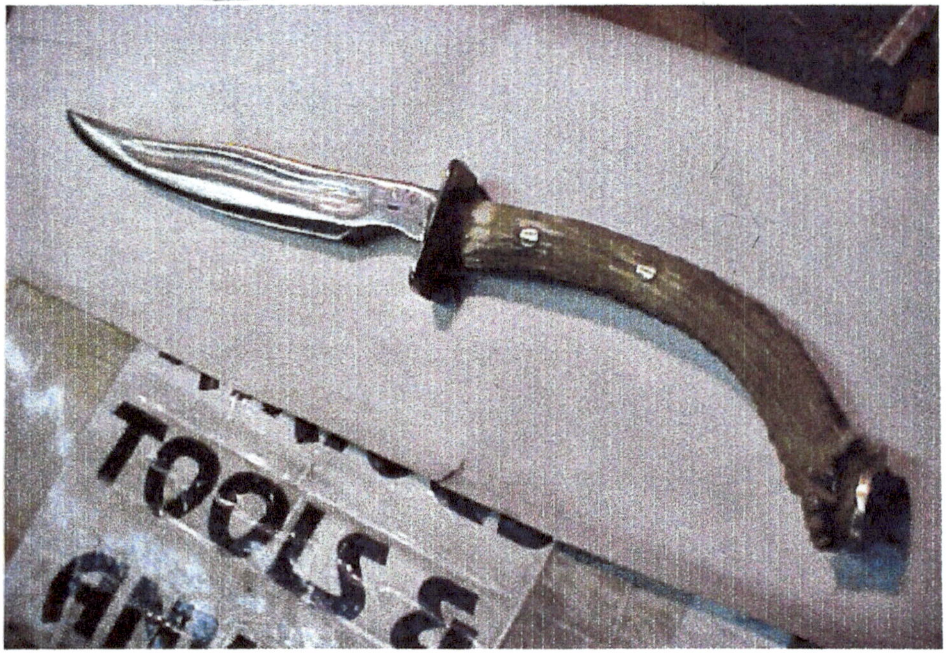

Picture 456
The project is now viewed for all contrast considerations of handle color. Antler is one of those mediums which can be lightened, darkened, polished, waxed, colored, smoothed, roughened, carved, burnished, or left alone…

Picture 457
At this stage our blade is etched in a mild acid solution. This is done outdoors with lots of ventilation. Time, temperature and solution concentration will produce any combination of visual affects here.

Picture 458
Give your etching solution time to show its results. Some steel combinations are more reactive than others. Wait and watch…

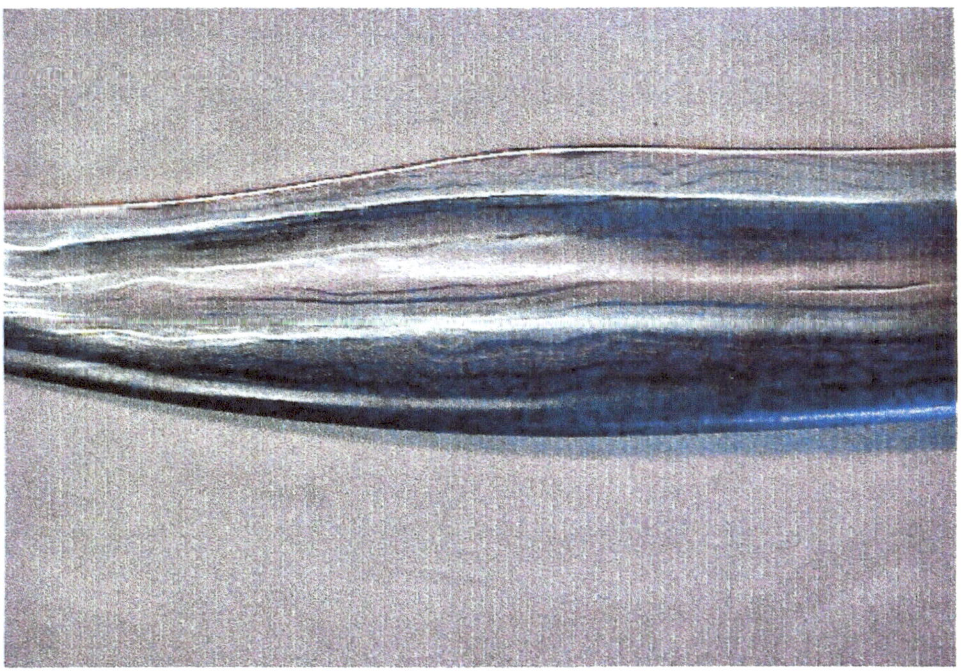

Picturc 459
And now we have what was in the search. Darkened areas are mild steel. Tool and carbon steel are less reactive to the acid etch for the soak duration time. The contrast now shows all of our welding lines and steel layers. This can be described as vibrant brilliance.

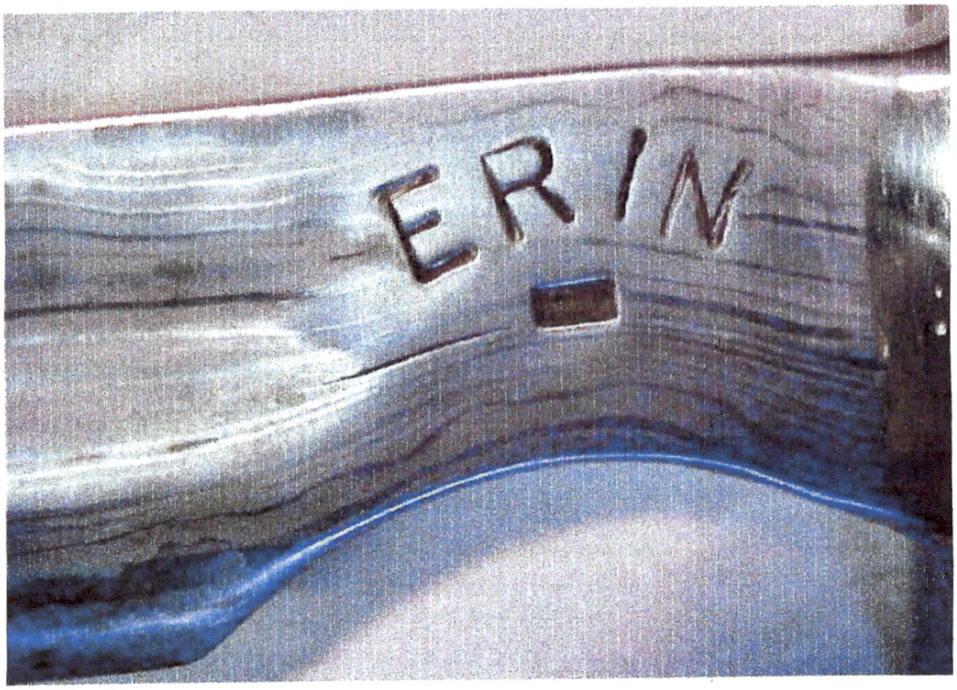

Picture 460
Each time one of these folded and welded blades is finished, it feels as though contact has been made with the smiths of old!

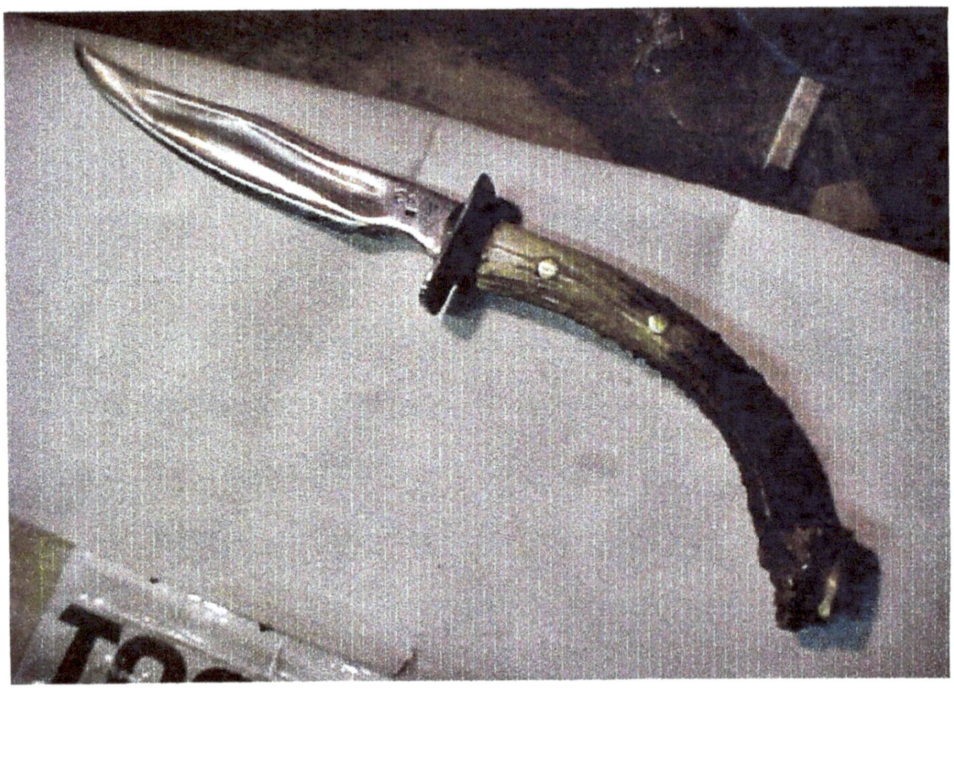

Picture 461
The finished product at last. The antler is hand sanded with soft 80 grit paper. Fire burnished to create contrast. Re-sanded to blend lighter colored areas to darker gradually. Then waxed with a saddle compound. Any excess is wiped away.

THE SPECIALTY SPRING

THE SPECIALTY SPRING

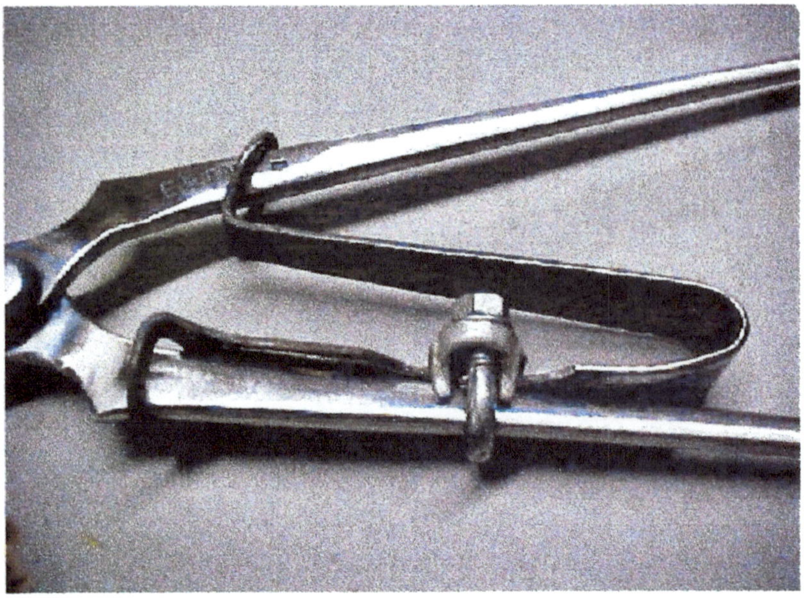

Picture 462
Pictured here are two pairs of tongs developed for someone needing the reins to return opened. The reason for their development was twofold. The farrier for which they were made had an accident which rendered his tong hand unable to open a conventional pair. The grip strength was at optimum only when the reins stopped at 2 inch width (outside width) while gripping the proper stock size. This is farther apart than usual. Shown are two different styles of spring loading for recoil. One provides for slightly more spring travel with uniform compression force throughout.

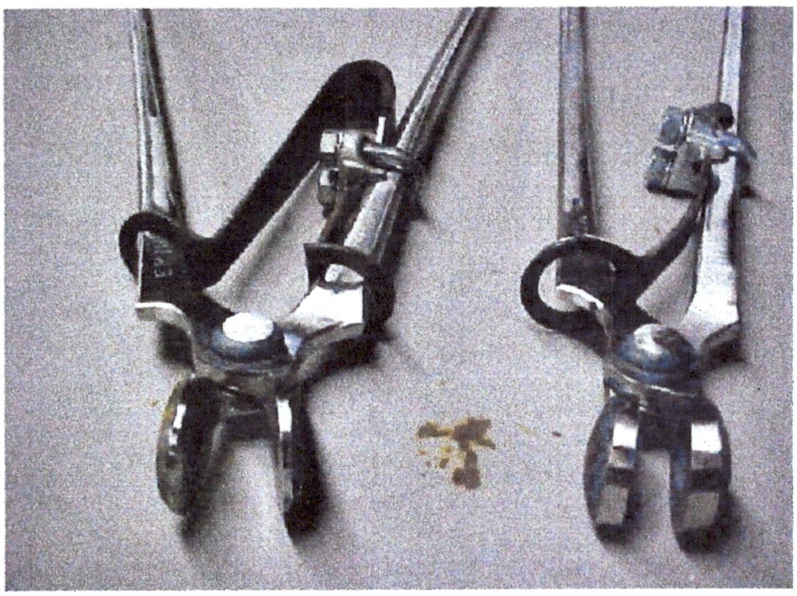

Picture 463
Jaw View

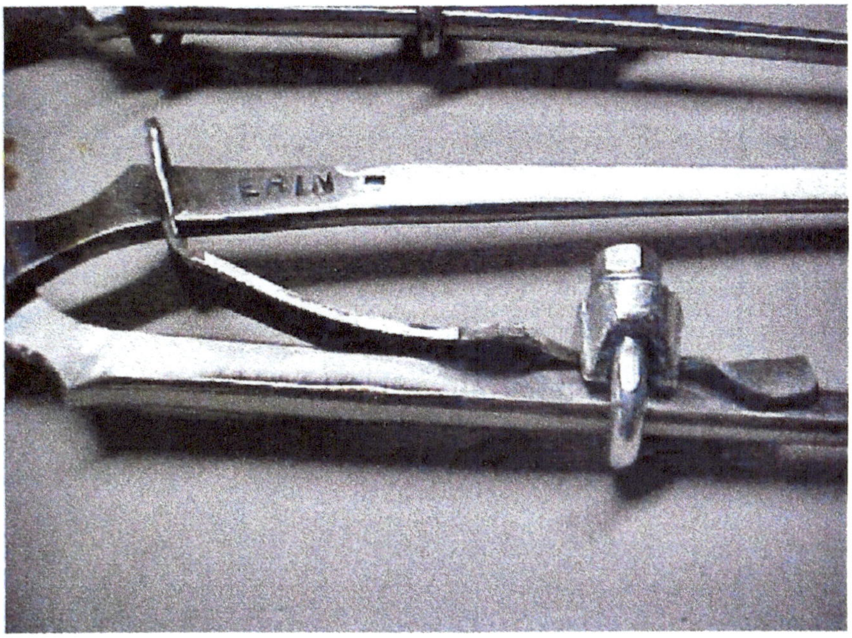

Picture 464
The mechanical advantages of our spring are much as a box leg vise of old. The pressure is applied closest to the riveted hinge.

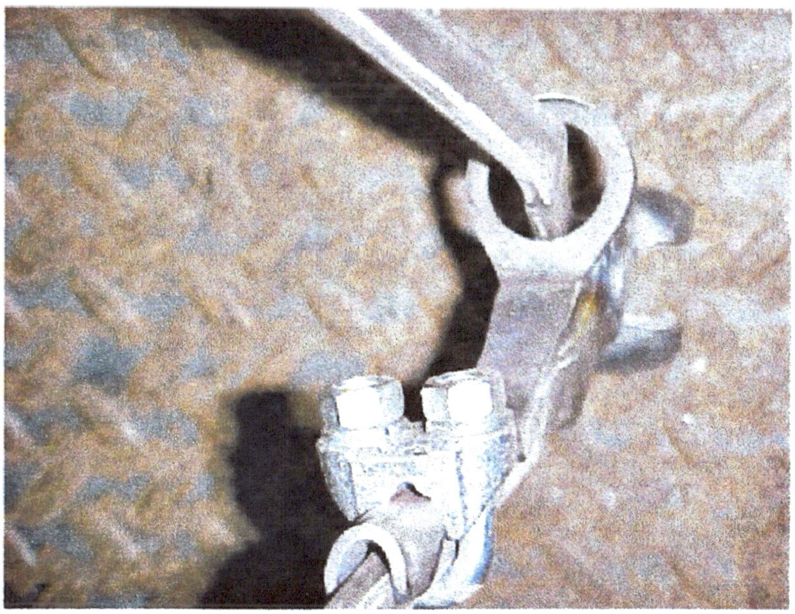

Picture 465
Rear view of the spring.

THE SIDE JAW TONG

THE SIDE JAW TONG

Picture 466
Side-jaw tong blanks from an old truck axle. These two pieces were from heavy stock.
If light stock is available, these will be easier to forge using our doubling over and welding methods, as rein drawing requirements are lessened. This particular pair has an "O" ring support rivet for load stability.

Picture 467

THE HAY HOOK

The Hay Hook

(for hand moving hay bales)

Stock dimensions per hook 5/16" x 5/8" x 29" A 36 of mild steel makes a very good size in weight and length

Forging Progression in proportion (not to scale)

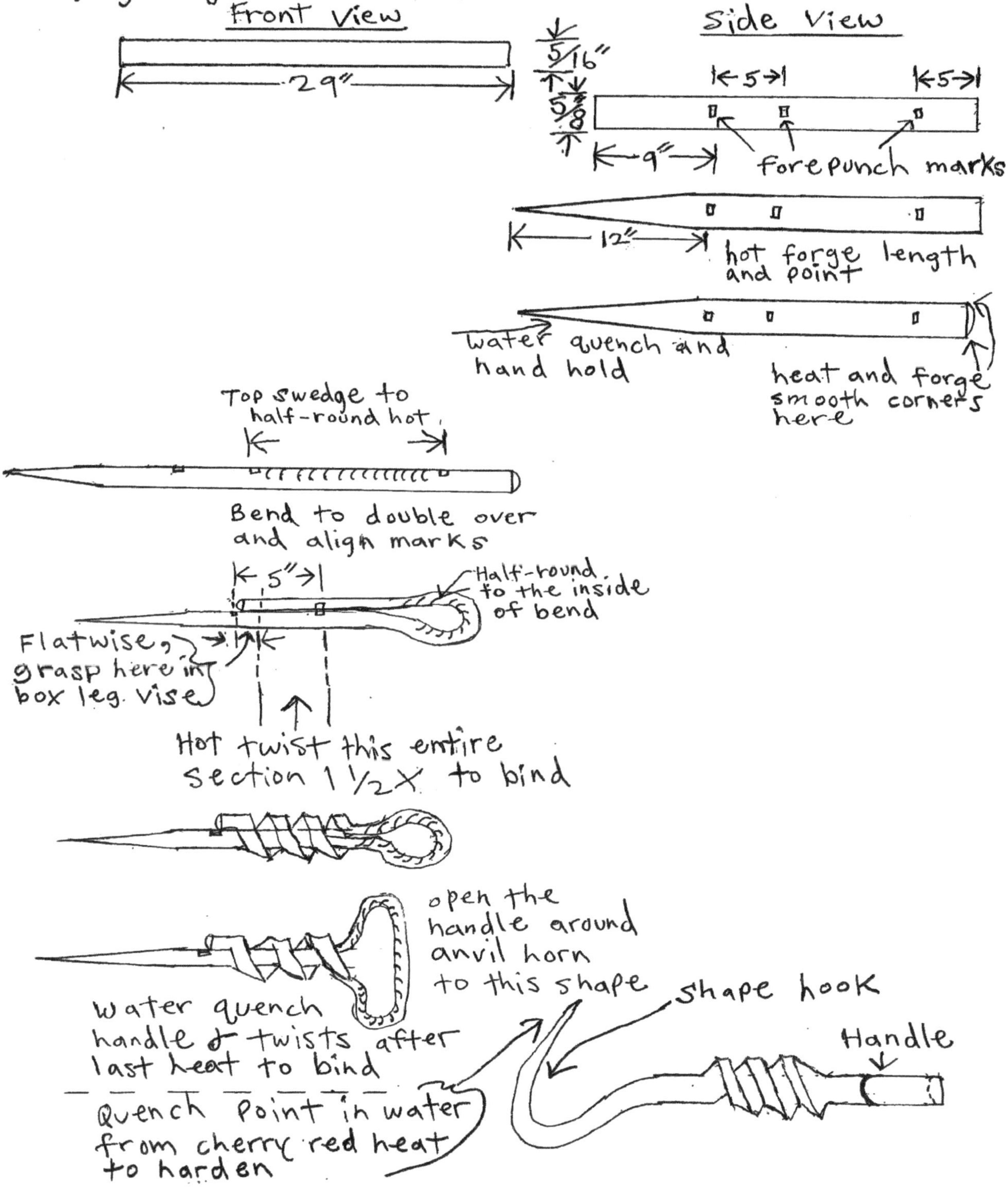

Picture 468
A close up of the handle twisting.

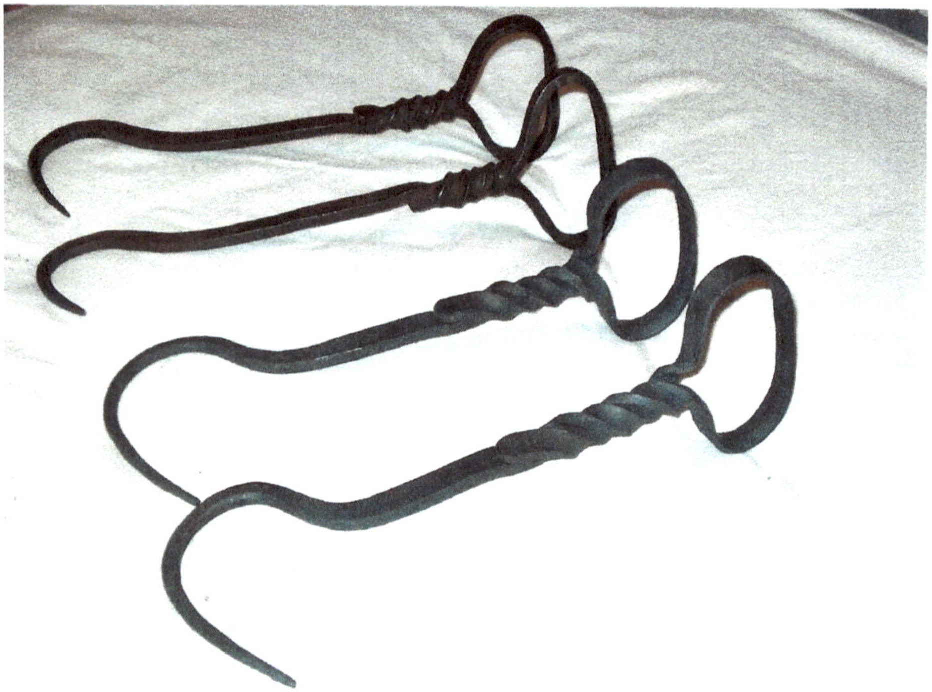

Picture 469
The finished product is now ready to be put in service tossing hay bales.

BOX JOINT HINGE TOOLS

Box Joint Hinge Tools

Applications: tong riveting, gates, latches, sturdy forged joinery

Special tools needed:- thick hot chisel (convex crowned striking and cutting surfaces) handled, tool steel (air hardening, S-7, H-13

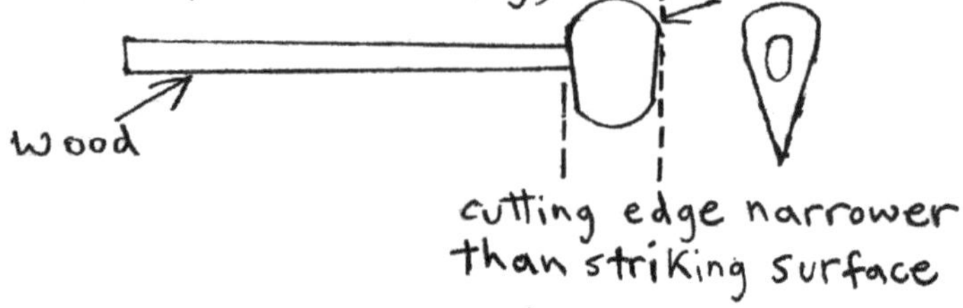

-bolster drift; handled to strike, side holed for rivet pin punching

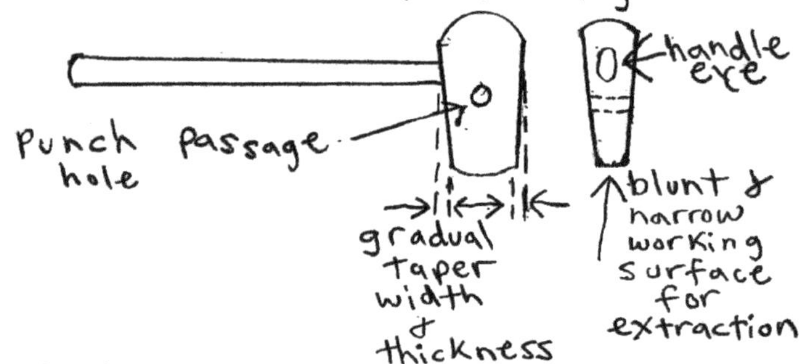

- rivet hole punches
  round: pivot type
  square: non-pivoting
  & rectangular

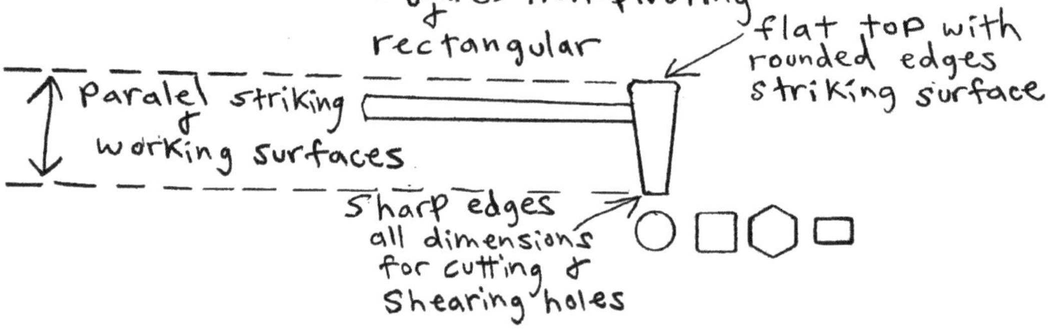

Have two or three of each tool ready for use. As one becomes hot, put it down to air cool. Continue work with a cold one. Cooler tools stay sharp longer.

THE BOX JOINT HINGED TONG

The Box Joint Hinged Tong

Materials: tool steel - sucker rod, jack hammer bit, breaker bar, coil spring are best
mild steel o.k. here

1 pair of tong { Double brace side 1¼" x 7" ; single brace side 1¼" x 5" } scale up or down as required

Round stock is good but double brace section will have to be squared up slightly to facilitate hot chisel splitting. Hexagonal is excellent to start with.

Double brace forging progression

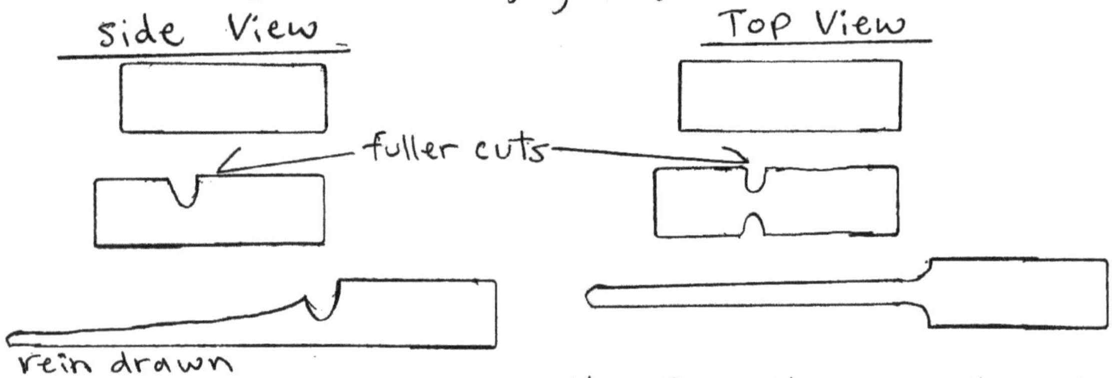

Hot chisel cut made completly from the top. The chisel is not as long the end slot desired. Walk the chisel. Start away from the smith with tool angled down. Finish a pass with the tool angled up. Draw the chisel toward the smith making many passes. Gradually gaining depth.

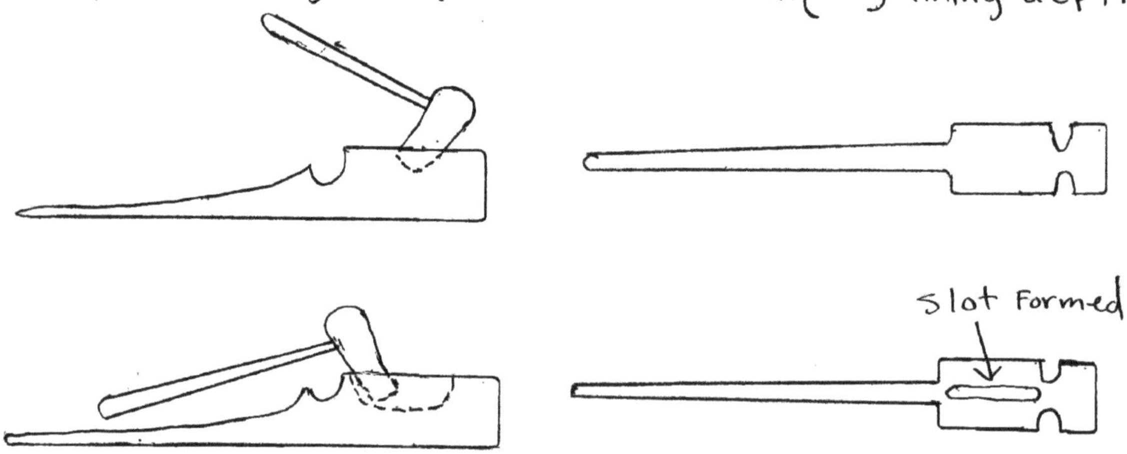

Side View | Top View

Passes continued until hot chisel reaches bottom. To make the final pass for slot opening, a copper plate is placed beneath for anvil and chisel protection.

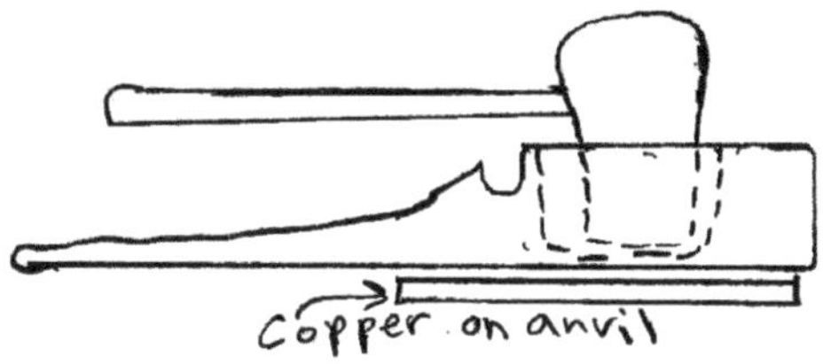

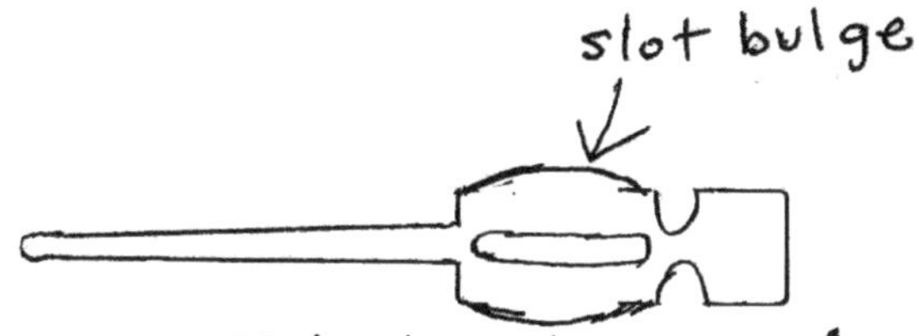

Slot through brace is complete.

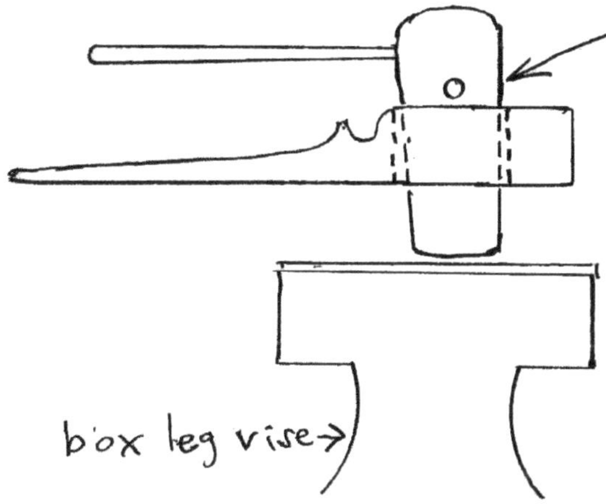

- Bolster drifting now begins atop a box leg vise opened only wide enough to accept the working edge.
- Brace sides are supported by the vice jaws.
- After a first slot drifting, the bolster is used as a mandril.
- While still in the slot, hammer the brace sides of the tong blank to become flat.
- The tong shape is completed before any rivet hole is punched or drilled. This includes jaw, boss and rein shoulder.
- Rein profile and width from its opposing mate can be adjusted lastly.

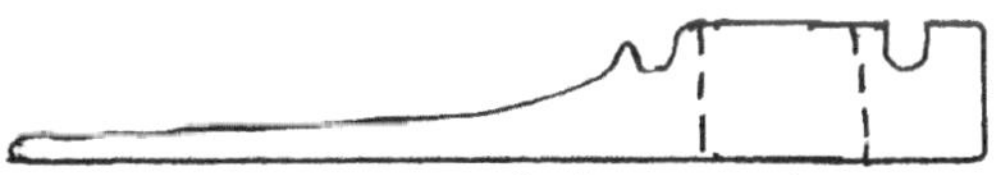

- This fuller cut is made nearer the forgings end to enable even pushing in the slot by hot chisel and later bolster drift.
- Defines tong jaw offset cleanly.

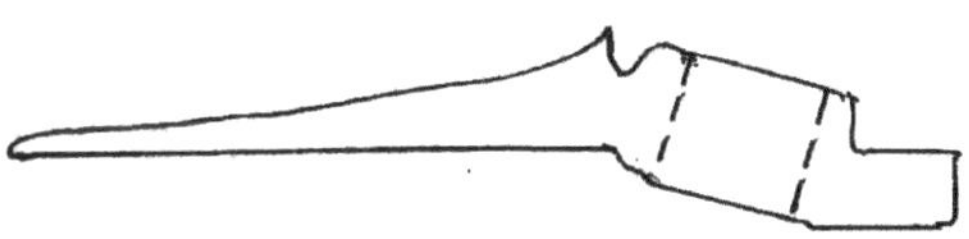

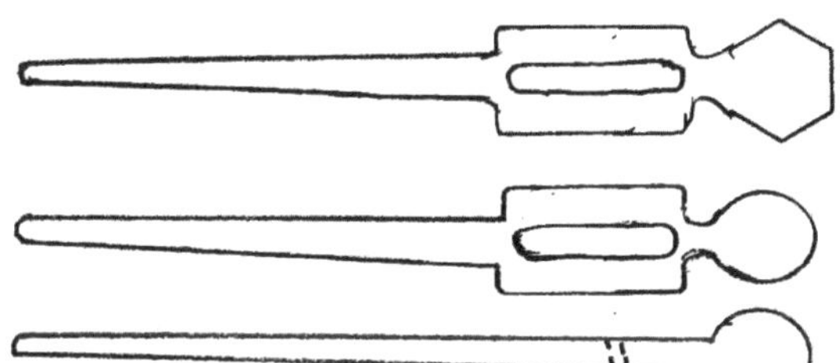

The opposing blank has no shoulder or jaw offset. Just forged and ground to fit through its mates slot.

RIVET HOLE PLACEMENT

Where Do We Place The Rivet Hole?

Tong blank Side View:

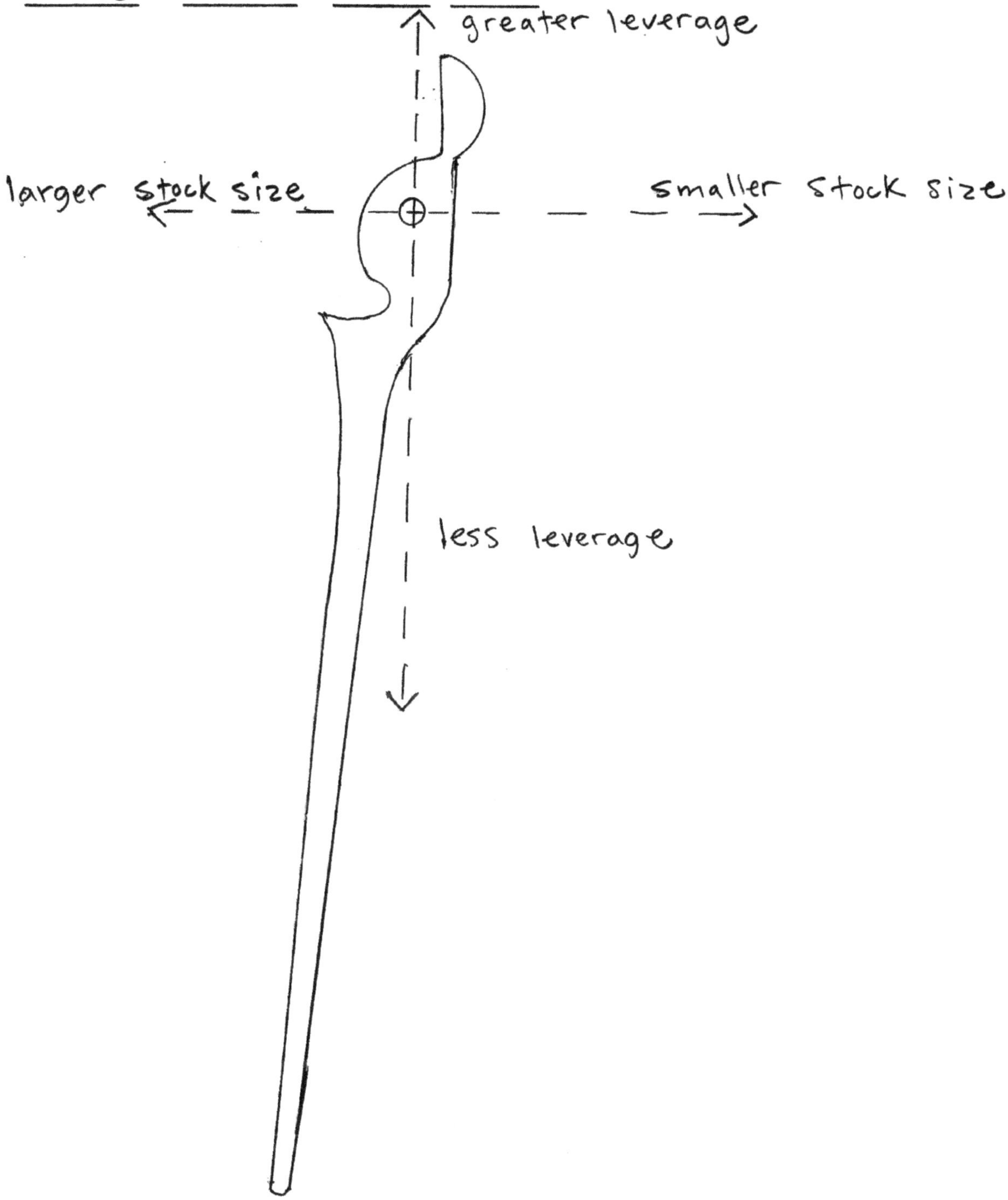

HOT PUNCHING RIVET HOLES
TO A
BOX JOINT TONG

Hot Punching Rivet Holes to a Box Joint Tong

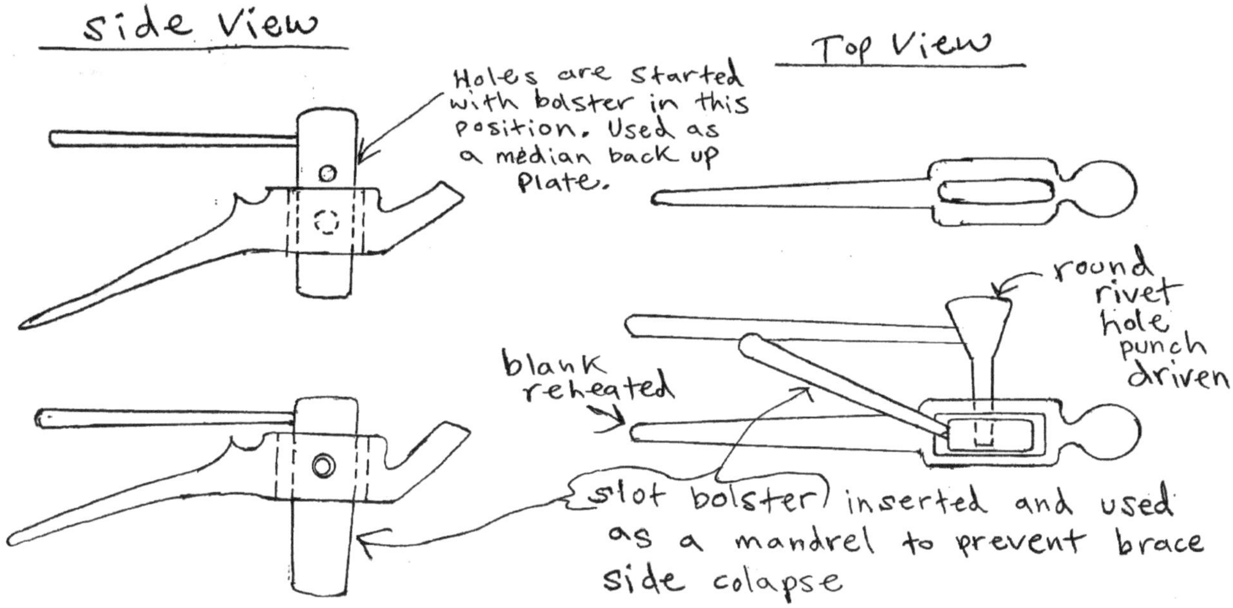

The brace rivet holes are started with our bolster blocking and backing up the area which will have the hole. Once the hole is started in the brace side, the bolster is driven through farther to center its passage hole under the brace side hole to be. Round hole punching then commences to push out a round button slug. The tong is hand hammered lightly to release the punch and bolster, should they become lodged. A light mark is left (just a halo's worth) inside our second brace after punching the first brace hole. A second, smaller diameter round punch is best here. This halo is then used to center our tools for punching the second brace hole. This helps all to remain true and correct when the finished product is in use. Our bolster and punch are extracted from the tong blank lengthwise along a narrowly opened box leg vise. Reheat the blank and continue punching the second brace after bolster insertion, outside → inside.

THE UNEVEN STOCK TONG

The Uneven Stock Tong

Most tongs are adjusted to hold specific sizes. During forging, dimensions can change which make a secure grasp very dificult. This tong design can accomadate many degrees of change and taper of flat stock.

Only one blank of this type is needed per pair of tongs.

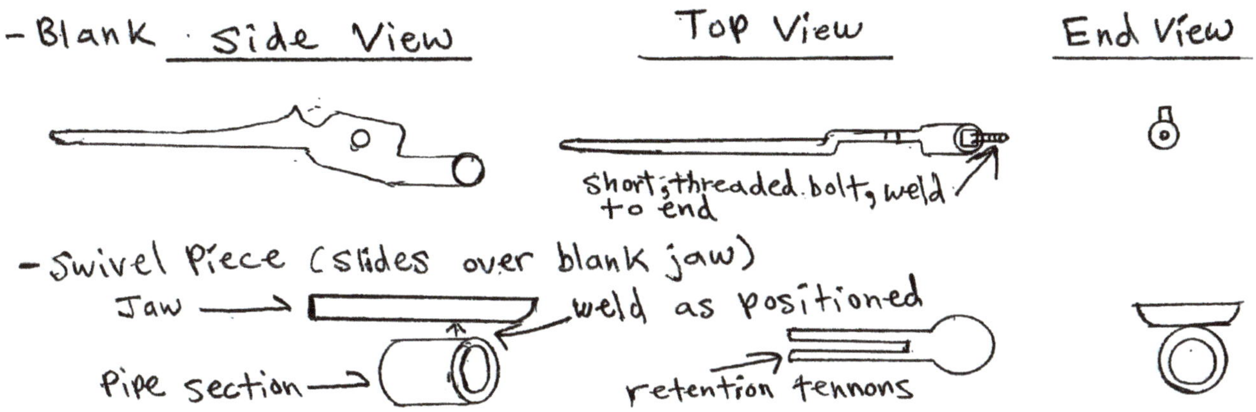

Swivel piece is held on blank jaw with a flat washer, lock washer and nut. The degree of movement depends only on the amount of distance between the retention tennons, Pipe fit over blank jaw is close tolerence, but free...

Picture 470

Picture 471

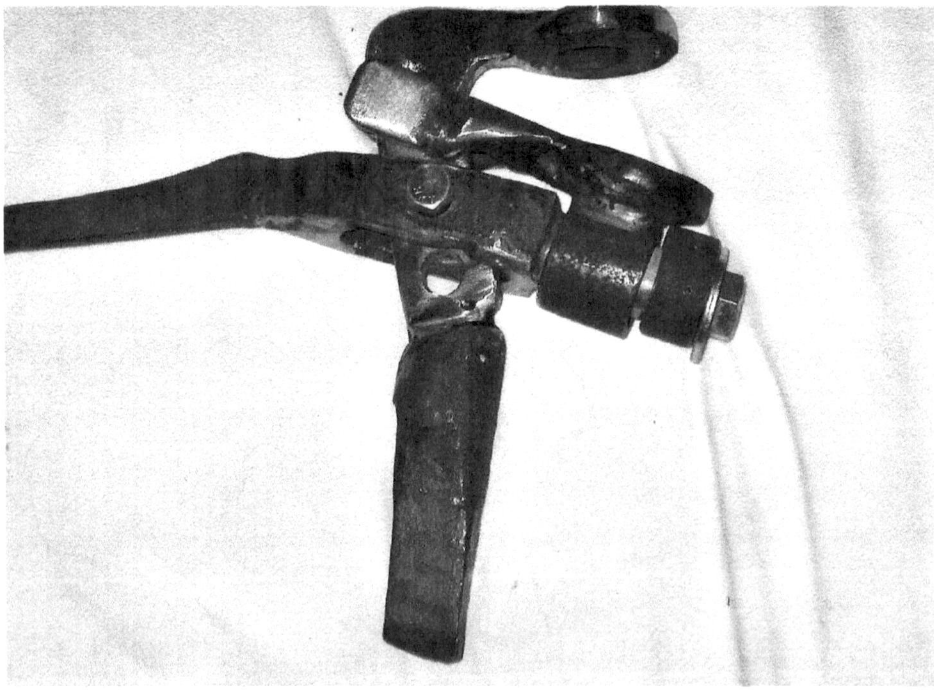

Picture 472

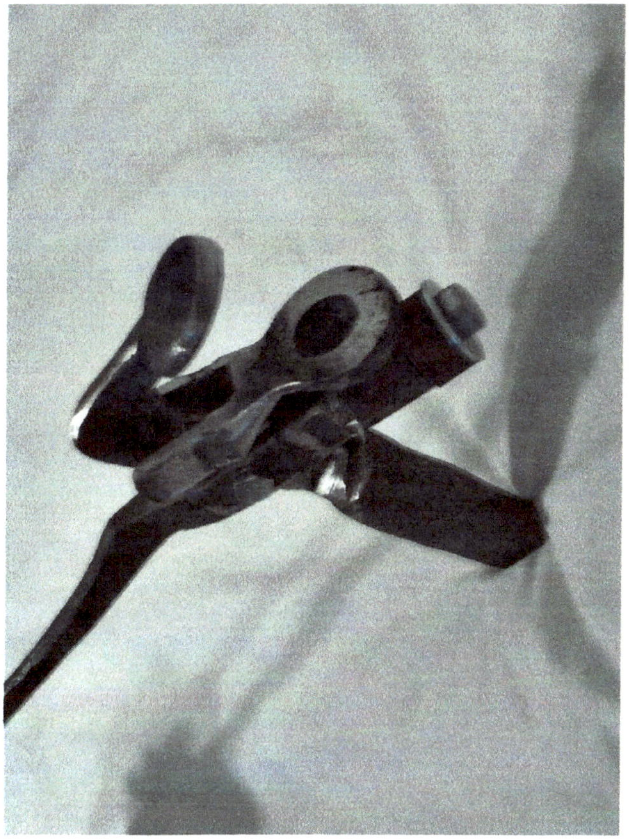

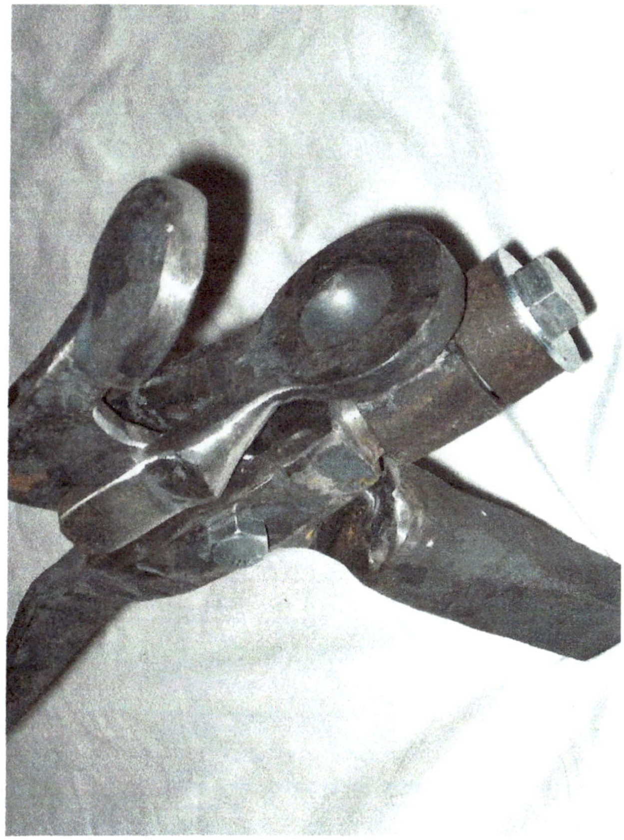

OUR CODES

Tools which have been forged and finished in our shop are coded for steel identification. In the early years each tool whether a hammer, set tool or even tong bore marks to denote the year in which it was forged. The tool dating is to help monitor their useful lives, longevity. Any return to the shop for repairs has proven less useful; as many may outlast their maker.

What has proven useful is the steel type code. As a billet of steel is cut, forged and near finished, a letter or number or combination of both is stamped into the product so that there is no mystery for us regarding the inherent steel composition. This is especially valuable for tools which must be heat treated. Tongs are not heat treated but ours are also coded. Dies and swedges are coded for steel composition, maker identification and heat treatment method.
The codes are as follows:

S - S7
H - H-13
C – coil spring (automotive)
TC – tractor coil (spring tooth harrow)
T – Titanium
R – rebar
10 – 1045 plow shear steel hot/cold rolled, new steel
W – diesel engine wrist pin
Hammer without mark – tractor track pin
Tong without mark – 4140, new steel
L – leaf spring
TR – train coil spring (locomotive)
M – mild steel

<u>Dies and swedges only</u>

W.T. – water hardened and tempered to straw
S.T. – salt brine hardened and tempered to straw

THE GREEN CHAIR

THE GREEN CHAIR

Everyone who is on their feet from sunup to well after sundown should have a good place to rest. When steel is pounded incessantly it's sure a comfort to look forward to daily. A befitting tribute to a green chair which has given at least 25 years of gracious service without complaint. Splendid behavior.

Picture 477

Our dog and cat keep a vigilant watch over the resting place, knowing certain that at some point their master will return.

THE CHAMPION

Picture 478

Pictured is a square ring made of 3/8 x 3/4 inch mild steel. The project was to find the proper length to cut, fire weld forge, and end with a 4 inch square creation. Something anyone can benefit from trying. This one was cut 17 ½ inches in the straight. A center punch mark was made flat-wise at 2, 6 ¼, 10 ½ and 14 ¾ inches respectively. A quarter inch was bumped by differential quenching into each center punch marked region while our steel was in the straight. These regions do best to become the box's square corners. The ends are scarfed and with a toe bend forging heat we hammer all into a round ring shape edge-wise-over the anvil horn. When the scarfed ends are brought together a ¾ inch overlap is ensued. After fluxing and fire welding we now have a full ring oval shape. The oval ring is now turned up on edge after heating and hand hammered on anvil face to square. The center punched and hence wider sections are driven down to become corners. We are going from straight to round to square in one hour to exact the 4 inch dimensions. When all goes reasonably well in a contest situation there may be a crown or winning belt buckle as champion blacksmith. Most often the personal satisfaction that comes from trying and participating is to win.

There are many means of tackling this project. Lengths can be cut to make all four corners a weld. "L" shapes can be forged to have two corners welded. Both of these methods can be cut and/or differentially quenched to bump up the thinner portions. A chief objective is to practice deductive reasoning rather than random assortment.

An experience has come to light here that nothing very good nor very bad lasts for very long. The nearby crowd of friends and family who cheer for you can feed the soul and spirit for a very long time.

IN THE SHOP

Picture 479

Picture 480

Pictures 479 and 480 – Shown on opposite page.

The display here is a compilation of numerous old tools and horse shoes which have created a personal challenge to improve upon. The rounding hammers were purchased from an antique dealer in Missouri who knew them to be 80-90 years old. They are now assumed to be well over 100 years.

A personal goal was to study the details bestowed upon them and learn what the smith's of old imparted, to create a quality forging. Making a hammer as good or better became the mountain to climb. We believe the mountain peak has been reached and still new challenges can be seen. Finding projects which are intriguing and challenge our skills from time to time will help a prospective smith, regardless of the years in service of the profession. They are nice to glance at once in a while between forging heats.

Picture 487 – Shown below.

A horse shoe board. The shoes and hoof pick possess forging elements that were essentia to complete a comprehensive farrier course of study at California Polytechnic State University.

Some of the forging skills included tending a coal fire to bend, draw, upset, flatten, widen, twist, punch, fire weld and braze, and hot rasping. Good horsemanship skills was and stil is a focus. Treasures to look back and improve upon…

To know where we are, we should remember where we came from. Everybody must start somewhere. Different backgrounds and experiences create a wondrous variety. This diversity has and will continue to provide new inclusions for the blacksmith's end product. We can endeavor to not only a size and magnitude of projects which suit our needs, but their usefulness and purpose. This adds to the realm, an aire which that smith can evolve. Present need is usually the matter of our inventions. Yet if a weakness is discovered why not delve into and practice this as an ambitious student. If that weakness can become strength, imagine the possibilities!

Picture 481

I BELIEVE

I Believe

That most individuals are good at something positive, there may have been dificulty in finding the pathway of principal achievement.

The magical qualities of steel and man must at some level find reality before all can find reason to marvel.

From shoeing horses to forging tools, there is such a thought process and practice that good enough is good enough.

Work loads accepted are relative to an individuals environment and moral quality standards.

If smithing was a sport, experience should be the referee.

Incompetence has just been sent to the penalty box.

The next athlete in should be knowledge.

Not to beg, borrow or steal this key team mate, just let the magic work.

That youth and vigor is disadvantaged against age and experience.

Time can heal, yet scars often leave a road map of damage severity for flesh or steel through callous and rust.

Some folks will forget more technique, preparation, planning, forging, grinding, machining, polishing, heat treatment and finishing lessons than can be remembered.

Iron is an element that awaits our organizing, it will not wait forever, but how long is a long time.

Breathing life into a cold inanimate object
is nothing short of magical.
If and when it then creats appeal to our peers
near and far, acceptance has been reached.
Should this acceptance rendor personal satisfaction,
our self esteem has been fed a gourmet meal.
When personal pride is not at stake, the work is
no doubt just a job.
Folly with arrogance in our persuit of glory creats
a substandard of quality finished product.
Though we all have to start somewhere, not all
compliment of equipment both within and without
our being may be found.
Many of the best have begun with the least.
In finding ways to learn, produce a product, sell
or trade goods or hard labors sweat for continued
practice and living, life can be passing by.
True as an arrow in flight, training does work, for
mortal sins against steel during forging can have
roots in amoral upbringing.
Wise should be a goal, not a birth right.
That wisdom comes in many forms.
How it is applied determines the variety we seek.
That wisdom comes from experience.
Experience forms mainly through poor judgement.
Better judgement appears over time as wise.

There can be truths about steel which have yet to be discovered.

Those with real answers, discoveries or partial successes will sooner find missing links by disclosure to other laymen and scientists.

Physical fortitude will only go so far; to consider and grasp true success mental aptitude must kick in some direction.

In the search, a place is found behind whit, cunning, experimentation, controversy, no surrender, and on the cerebral spectrums forefront, the obvious.

It can be more important to track our progress than to focus on the steps it to get us there.

Most useful attributes result when slowly looking ahead.

High moral fiber can be measured as much for deeds attempted as for those completed. Where an honest try runs its coarse, we each find moral fortitude.

Moral solitude does not always equal fiscal plentitude,

Balance of tasks helps blend a days chores.

The quality of smithing life may have a proportion of rate versus duration as do angles of physics in rise over run.

The more revenue per hour/week/month/year the greater our net worth, as society places much value.

Output is not always a proportionate virtue, for greater productivity per time can make no less valuable our skills and hence net worth.

Even a blind squirrel finds an acorn once in a while.
We may only begin to understand parts of our projects at different stages of a career.
In savoring those parts which help solve some misunderstood elements.
Observing, remembering and comparing minute results to true success adds background.
Errors can also build a library...
Medical doctors and veterinarians will tend to flesh and bones for a work medium.
In aiding the repair and knit of wounds upon the old and young.
Repairs of cosmetic or dire functional necesity meet with equal protocol.
The spirit of each patient or consult may need healing even for the healthy.
a smith works to reconstruct ferous materials as their medium.
The scalpal is hammer and hot chisel.
The knit is a weld, a braze or a rivet.
Most creatures that occur in nature do not have roots and wings.
Smiths can be called upon to create parts of items which fly or remain anchored firmly to the ground, I believe.

CONCLUSION

CONCLUSION

As the ideas conveyed evolve into actual projects, a breath should be taken. Remember that these ideas are just that. The smith must learn to think for oneself. To make sense of a forging by asking mental questions. How will the joinery, if any, best be constructed? Sometimes, how can a project be fabricated? There can be many possible methods but only a few practical means of getting it done. The path of walking/running at a new project is the venue to be practiced. Learning how to formulate an idea and bring it to life. Taking the intangible and bringing it to reality. This path in reference has more to do with an individual's thought process. The practical and logical versus astrophysical. Do items of artistic design put more fuel in our cerebral tank or is it projects of functional means? This can be the mundane versus the exceedingly tedious spectrum. For my own particular interests, tools to be forged get the innovative process from a walk to a brisk trot on the cerebral path. This path has been well worn. It is this initiation of each smith's thought process which can help a learning curve. Learning how to look for and then run ragged the creativity road. From an ideas inception to conclusion we each can learn to isolate parts of a project and tie in other supportive elements. Many of our attempts in blacksmithing projects have four elements which get us on track:

1. An idea
2. A beginning
3. A Middle
4. An End

We, therefore, mentally need to learn how to find projects which appeal to our needs, whether newly formulated or pre-existing. They can be for profit or for practicality, dreamt up and immaculately conceived. Next, engaging the desire to pull together mentally any/all constraints needed to commence your undertaking. Right here we are teaching ourselves the first two of my blacksmithing elements!

Finding and becoming accustomed to traveling this project element path will go from a seldom traveled road to as worn as our kitchen to the dinner table.

My best,

Erin A. Simmons

Jasmine, Erin & Rabbit rest

ISBN: 0-9762739-0-X (Volume I)

www.ingramcontent.com/pod-product-compliance
Lightning Source LLC
LaVergne TN
LVHW070945210926
848482LV00002B/94